I0053576

Space Nuclear Propulsion and Power

Book 2

Nuclear Thermal Propulsion Systems

David Buden

Space Nuclear Propulsion and Power Book 2: Nuclear Thermal Propulsion Systems

Polaris Books
11111 W. 8th Ave., Unit A
Lakewood, CO 80215

www.polarisbooks.net

Copyright © 2011 by David Buden
All rights reserved, including the right of reproduction in whole or in part in any form.

Manufactured in the United States of America

First Polaris Books Edition 2011

ISBN 978-0-9741443-3-7
Library of Congress Control Number: 2011911293

Cover image by Pat Rawlings of Science Applications International Corporation for NASA
Book and cover design by Alyssa Piccinni

CONTENTS

Foreword

Propulsion and power are defining technologies that enable man to perform more demanding space missions. In 1985, Orbit Book Company, Inc. published *Space Nuclear Power* by Joseph A. Angelo, Jr. and David Buden. Since that time, much has happened in the field of space nuclear systems. This series of three books on space nuclear power and propulsion:

> Book 1: *Space Nuclear Radioisotope Systems*
> Book 2: *Nuclear Thermal Propulsion Systems*
> Book 3: *Space Nuclear Fission Electric Power Systems*

brings together a summary of all of the developments in space nuclear systems.

Dr. Mohamed S. El-Genk organized the Symposium on Space Nuclear Power and Propulsion in Albuquerque, New Mexico from 1983 thru 2008. Dr. El-Genk is Regents' Professor, Chemical, Nuclear, and Mechanical Engineering, and Director of the Institute of Space Nuclear Power Systems at the University of New Mexico. A debt of gratitude is owed to Dr. El-Genk for his tireless efforts in bringing together the principle investigators in the space nuclear field and publishing symposium proceedings for each of the yearly meetings. The proceedings were extremely helpful in preparing this book.

Dr. Gary L. Bennett offered invaluable insight and assistance in preparing the manuscript of this book. He worked for NASA and the US Department of Energy (DoE) on advanced space power and propulsion systems. Dr. Bennett started his space nuclear career on the nuclear rocket program (NERVA) at NASA Lewis Research Center. Subsequently, he has held key positions in DoE's space radioisotope power program, including serving as the flight safety manager and Director of Safety and Nuclear Operations for radioisotope power sources used on Voyager 1 and 2 spacecraft to Jupiter, Saturn, Uranus, Neptune and beyond; Lincoln Laboratory's LES 8 and 9 communications satellites; and the Galileo mission to Jupiter. At NASA, he was the Manager of Advanced Space Power Systems in the transportation division of the Office of Advanced Concepts and Technology. He managed, among other things, fission nuclear propulsion activities.

Overview

Key elements in the exploration and exploitation of space are reliable and compact propulsion and power systems. Without adequate power and propulsion, space missions are severely limited. Nuclear powered systems are a key technology in meeting past and future propulsion and power needs. The current generation of nuclear systems can be organized in terms of radioisotope and fission nuclear reactor heat sources. Fission nuclear systems can be subdivided into power and propulsion applications. These are the bases for this series of three books:

> Book 1: *Space Nuclear Radioisotope Systems*
> Book 2: *Nuclear Thermal Propulsion Systems*
> Book 3: *Space Nuclear Fission Electric Power Systems*

It is vital to understand past activities in developing future programs to meet the challenges that lie ahead. In order to meet the needs and interest of individual readers, each book in this series is designed to stand-alone.

Solar energy has been the cornerstone of power systems for near-Earth missions. However, the solar flux drops as the inverse square of the distance from the Sun (See Fig. 1).[1] At Jupiter, for instance, the solar flux is only 4% of that at Earth. Solar arrays become too large and weigh too much at such distances. Also, surface operations in sun starved or shadowed areas at nearer distance cannot use solar technology. Near the Sun high temperatures also limits the use of solar technology.

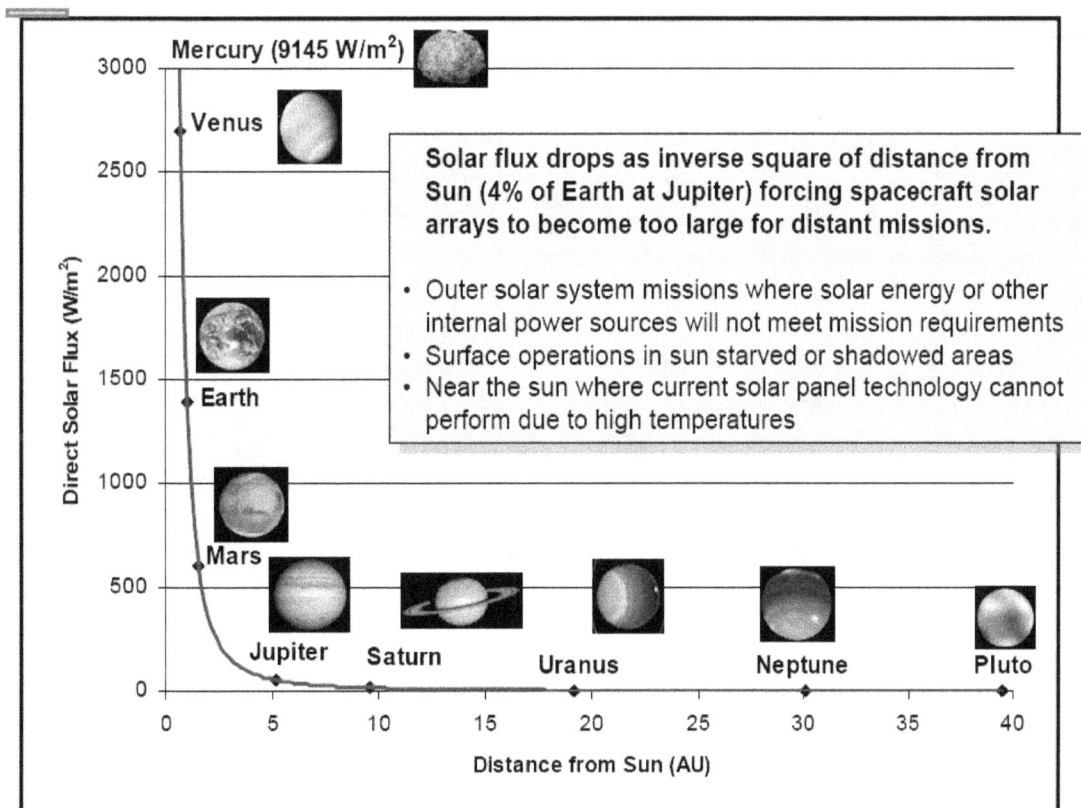

Fig. 1 Limited solar flux as distances increase from the Sun enables the need for nuclear space power systems.

To date, space propulsion has relied mainly on chemical fueled rockets. As missions with increasingly larger payloads are contemplated, the size of chemical rockets becomes unwieldy. Other more efficient propulsion systems, such as nuclear fission rockets, will be required.

Nuclear power has been used on many space missions by the United States in support of both civilian and military programs. These have taken the form of using the thermal energy from the decay of radioisotopes and converting this energy to electric power. Radioisotope power systems have proven to be highly reliable, operating for many years and in severe environments (e.g., trapped radiation belts, surface of Mars, moons of the outer planets) that make solar alternatives of limited use or unusable. Also, radioactive decay heat has been used to maintain temperatures in spacecraft at acceptable conditions for other components.

Radioisotope power systems are limited in power levels to a few kilowatts by the cost and availability of suitable radioisotope thermal heat sources. Nuclear reactors can provide tens-to-hundreds and even megawatts of power in future power systems. Extensive development works has already been performed on reactor-powered systems with the United States having flown one system in space. To meet the requirements of the more ambitious missions of the future, nuclear fission power will be a necessity.

Interest in nuclear rockets has centered on manned flights to Mars. The demands of such missions requires rockets that are several times more powerful than the chemical rockets in use today. Nuclear fission rockets have been extensively developed for this purpose. However, none of these developments have reached flight status.

The advantages of space nuclear systems can be summarized as: compact size; low to moderate mass; long operating lifetimes; operation in extremely hostile environments; operation independent of the distance from the Sun or of the orientation to the Sun; and high system reliability and autonomy.[2, 3, 4] In fact, as power requirements approach hundreds of kilowatts and megawatts, nuclear energy appears to be the only realistic power option (see Fig. 2).[5]

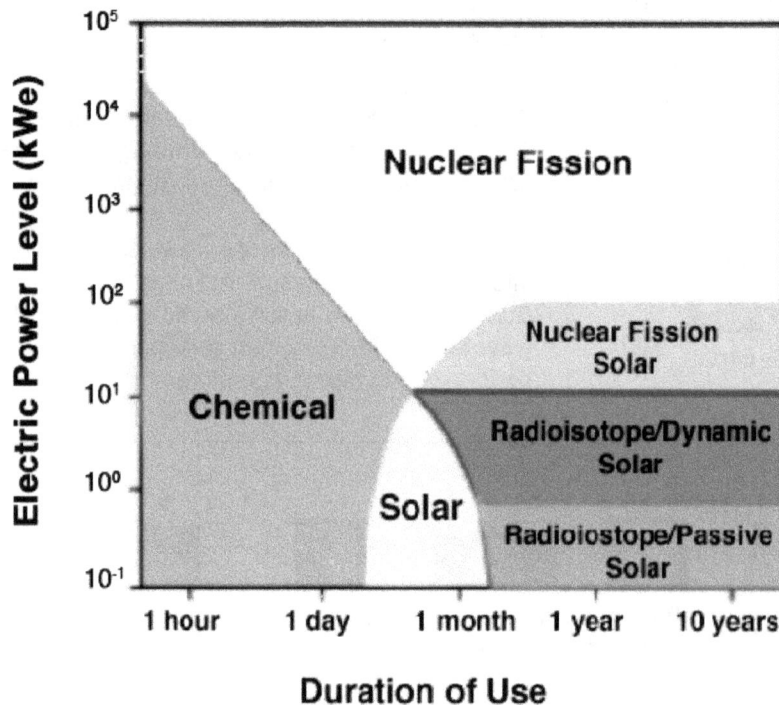

Fig. 2. Regimes of possible space power applicability.

The building blocks for space nuclear electric power system are depicted in Fig. 3. Radioisotope decay heat or the thermal energy released in nuclear fission can be converted to electrical using power generation equipment. These can take the form of either static electrical conversion elements that have no moving parts (e.g., thermoelectric or thermionic) or dynamic conversion elements (e.g., the Rankine, Brayton or Stirling cycle). The options for nuclear energy heat sources and companion power generation subsystems are summarized in Fig. 4. Radioisotope and reactor power systems are further classified in Fig. 5.[6]

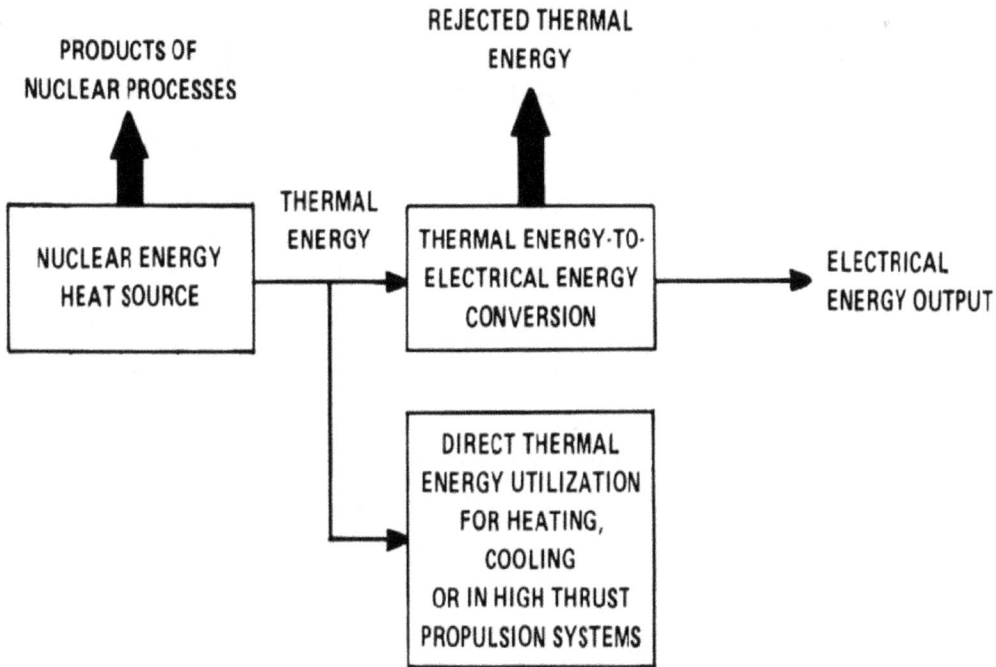

Fig. 3. Generic space nuclear power systems. *Courtesy of Los Alamos National Laboratory.*

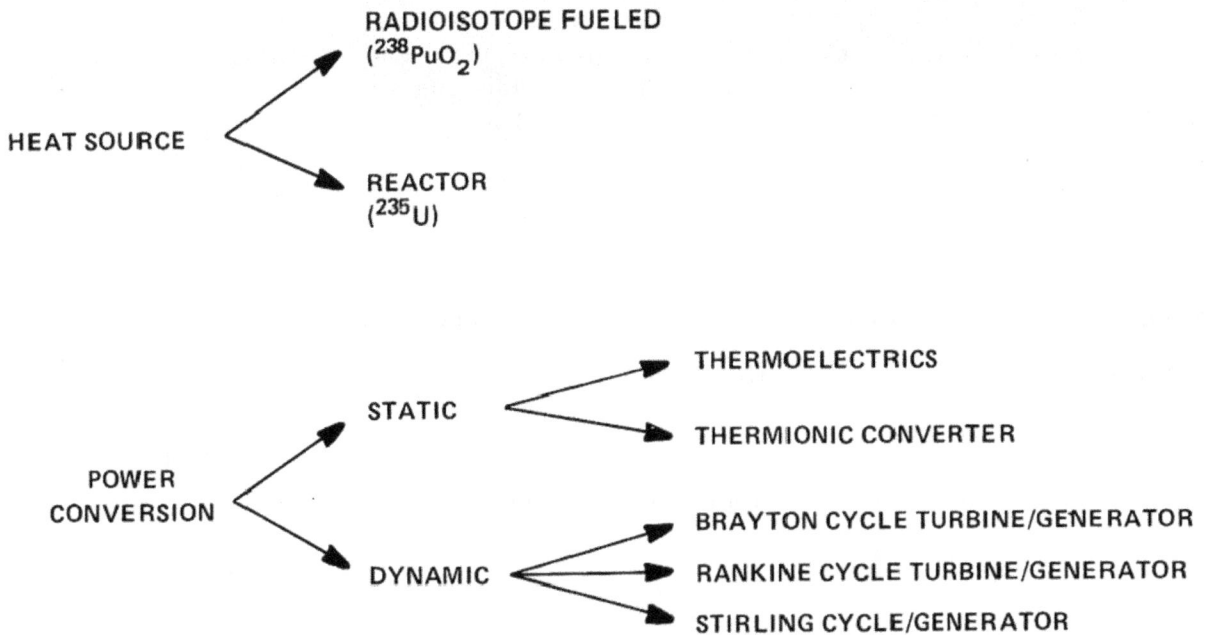

Fig. 4. Options covered in space nuclear programs. *Courtesy of Los Alamos National Laboratory*

Nuclear Power System	Electric Power Range (Module Size)	Power Conversion
Radioisotope Thermoelectric Generator (RTG)	Up to 500 We	Static: Thermoelectric
Radioisotope Dynamic Conversion Generator	0.5 to 10 kWe	Dynamics: Brayton Rankine Stirling
Reactor Systems Heat Pipe Solid Core Thermionics	10 kWe to 1,000 kWe	Static: Thermoelectrics Thermionics Dynamics: Brayton Rankine Stirling
Reactor Systems Heat Pipe Solid Core	1 to 10 MWe	Dynamics: Brayton Rankine Stirling
Reactor Systems Solid Core Pellet Bed Fluidized Bed Gaseous Core	10 to 100 MWe	Brayton Cycle (Open Loop) Stirling MHD

Fig. 5. Classification of nuclear power system types being considered for space applications. *Courtesy of Los Alamos National Laboratory.*

Since 1961, the United States has launched twenty-seven National Aeronautics and Space Administration (NASA) and military space systems that derived all or part of their power requirements from nuclear energy sources. These systems are summarized in Table 1; some illustrations of nuclear powered spacecraft are shown in Fig. 6. As can be seen in this table, all but one of the previous missions used plutonium-238 as the fuel in various radioisotope thermoelectric generator (RTG) systems. The SNAP-10A was a compact nuclear fission reactor that used fully enriched uranium-235 as the fuel. The acronym SNAP stands for Systems for Nuclear Auxiliary Power, with the odd-numbered units representing radioisotope heat sources and the even-numbered units nuclear reactors.

Table 1. Summary of space nuclear power systems launched by the United States.[7, 8, 9]

Power Source	Spacecraft	Mission Type	Launch Date	Status
SNAP-3B7	Transit 4A	Navigational	6/29/1961	RTG operated for 15 years. Satellite now shutdown but operational.
SNAP-3B8	Transit 4B	Navigational	11/15/1961	RTG operated for 9 years. Satellite operated periodically after 1962 high altitude test. Last reported signal in 1971.
SNAP-9A	Transit 5-BN-1	Navigational	9/28/1963	RTG operated as planned. Non-RTG electrical problems on satellite caused satellite to fail after 9 months.

Power Source	Spacecraft	Mission Type	Launch Date	Status
SNAP-9A	Transit 5-BN-2	Navigational	12/5/1963	RTG operated for over 6 years. Satellite lost ability to navigate after 1.5 years.
SNAP-9A	Transit 5-BN-3	Navigational	4/21/1964	Mission was aborted because of launch vehicle failure. RTG burned up on re-entry as designed.
SNAP-10A REACTOR	SNAPSHOT	Experimental	4/3/1965	Successfully achieved orbit. Operated 43 days above 785 K.
SNAP-19B2	Nimbus-B-1	Meteorological	5/18/1968	Mission was aborted because of range safety destruct. RTG heat sources recovered and recycled.
SNAP-19B3	Nimbus III	Meteorological	4/14/1969	RTGs operated for over 2.5 years.
ALRH	Apollo 11	Lunar Surface	7/14/1969	Radioisotope heater units for seismic experimental package. Station was shut down 8/3/1969.
SNAP-27	Apollo 12	Lunar Surface	11/14/1969	RTG operated for about 8 years until station was shut down.
SNAP-27	Apollo 13	Lunar Surface	4/11/1970	Mission aborted on the way to the moon. RTG re-entered earth's atmosphere and landed in South Pacific Ocean. No radiation was released.
SNAP-27	Apollo 14	Lunar Surface	1/31/1971	RTG operated for over 6.5 years until station was shut down.
SNAP-27	Apollo 15	Lunar Surface	7/26/1971	RTG operated for over 6 years until station was shut down.
SNAP-19	Pioneer 10	Planetary	3/2/1972	RTGs still operating. Spacecraft successfully operated to Jupiter and is now beyond orbit of Pluto.
SNAP-27	Apollo 16	Lunar Surface	4/16/1972	RTG operated for about 5.5 years until station was shut down.
Transit-RTG	"Transit" (Triad-01-1x)	Navigational	9/2/1972	RTG still operating.
SNAP-27	Apollo 17	Lunar Surface	12/7/1972	RTG operated for almost 5 years until station was shutdown.
SNAP-19	Pioneer 11	Planetary	4/5/1973	RTGs operating. Spacecraft successfully operated to Jupiter, Saturn, and beyond.
SNAP-19	Viking I	Mars Surface	8/20/1975	RTGs operated for over 6 years until lander was shut down.
SNAP-19	Viking 2	Mars Surface	9/9/1975	RTGs operated for over 4 years until relay link was lost.
MHW-RTG	LES 8	Communications	3/14/1976	RTGs still operating.
MHW-RTG	LES 9	Communications	3/14/1976	RTGs still operating.
MHW-RTG	Voyager 2	Planetary	8/20/1977	RTGs still operating. Spacecraft successfully operated to Jupiter, Saturn, Uranus, Neptune, and beyond.
MHW-RTG	Voyager 1	Planetary	9/5/1977	RTGs still operating in 2010. Spacecraft successfully operated to Jupiter, Saturn, and in the heliosphere.

Power Source	Spacecraft	Mission Type	Launch Date	Status
GPHS-RTG	Galileo	Planetary	10/8/1989	Completed 34 orbits of Jupiter. Mission ended 21 Sept. 2003 when spacecraft was plunged into Jupiter's atmosphere.
GPHS-RTG	Ulysses	Planetary/Solar	10/6/1990	Spacecraft completed nearly three complete orbits of Sun. Operated more than 18 y, shutdown 6/30/2009.
GPHS-RTG	Cassini	Planetary	10/15/1997	RTGs still operating. Spacecraft orbiting Saturn.
GPHS-RTG	New Horizons	Planetary	1/19/2006	RTG still operating. Spacecraft en route to Pluto
MMRTG	Mars Science Laboratory	Mars Surface	Pending 2011	

The Russians have also have made extensive use of nuclear power systems in space. They launched two radioisotope systems in 1965. In the time period between 1971-1988, they launched some 35 nuclear reactor systems (see Table 2).

Table 2. USSR space power flight experience.[10]

TYPE	NAME	MISSION	NUMBER OF MISSIONS	LAUNCH DATES	STATUS	FAILURES
RTG		NAVIGATION SATELLITES	2	9/65	IN ORBIT	NONE KNOWN
RADIO-ISOTOPE HEATER UNIT		LUNAR ROVERS	4	9/69 TO 1/73	TWO SHUTDOWN ON MOON	TWO REENTRIES AFTER UPPER STAGE MALFUNCTIONS (1969)
REACTOR	RORSAT	OCEAN SURVEILLANCE	35	12/67 TO 3/88	31 SHUTDOWN AND BOOSTED TO HIGH ORBITS	- TWO LAUNCH ABORTS (1969, 1973) - TWO REENTRIES AFTER BOOST FAILURE (1977, 1982)
REACTOR	TOPAZ	OCEAN SURVEILLANCE	2	1/87 TO 10/87	SHUTDOWN AND BOOSTED TO HIGH ORBITS	NONE KNOWN

Safety has been and continues to be a key element in the development and deployment of space nuclear systems. The prevention of significant radiological risk to the Earth's population or to the terrestrial environment are guiding policies in all phases associated with space nuclear systems.[11] For radioisotope heat sources, this aerospace nuclear safety policy essentially consists of providing containment that is not prejudiced under any circumstance including launch accidents, reentry, or impact on land or water. For nuclear reactors, the safety mechanism consists of maintaining sub criticality under all conditions, normal and otherwise, in the Earth's atmosphere or on the Earth's surface. After the reactor has experienced power operation in space, the reactor will be prevented from reentering the terrestrial biosphere until the fission products and other radioactive materials no longer represent a radiological risk. Reactor operation is generally limited to orbits that have a lifetime in excess of three hundred years to support this safety approach.

Apollo (1969 - 1972)

Voyager (1977)

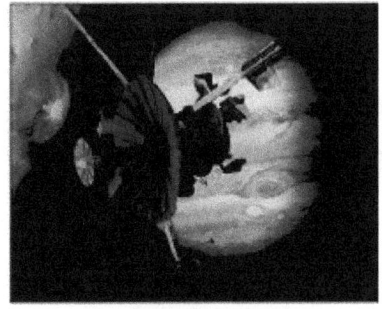
Galileo (1989)

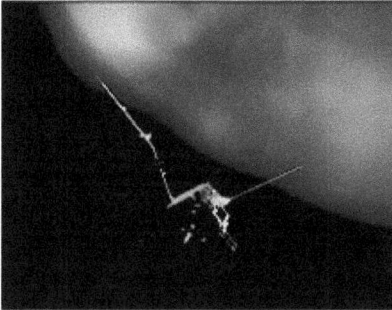
Ulysses (1990)

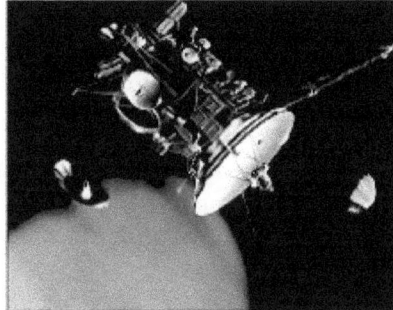
Cassini (1997)

New Horizons (2006)

Fig. 6. Illustrations of missions using nuclear power. All missions operated far beyond their design life times. *Courtesy of Dennis Miotia.*

Chapter 1

Fundaments of Nuclear Propulsion

Thermal Nuclear Rockets?

Chemical rocket propulsion has been the main stay of space propulsion; especially in respect to manned operations. These have served well for both the Apollo lunar program and the Manned Space Station. However, the performance of chemical rockets is limited. This limitation makes more demanding missions, such as manned trips to Mars, very unwieldy.

Chemical rocket engines produce thrust by the expulsion of high-speed fluid exhaust produced by the combustion of solid or liquid propellants. These propellants consist of fuel and oxidizer components that are burned within a combustion chamber. Therefore, they can function in a space environment. The fluid exhaust is then passed through a nozzle that uses the heat energy of the gas to accelerate the fluid exhaust to very high speeds. The reaction to this pushes the spacecraft in the opposite direction from the exhaust and thus propels the spacecraft forward. In rocket engines, high temperatures and pressures are highly desirable for good performance.

Rocket engines operate according to the basic principles expressed in Newton's third law of motion: for every action there is an equal and opposite reaction. In a chemical rocket, hot gases are created by chemical combustion; in a nuclear rocket heating the propellant in a nuclear reactor creates hot gas.   In either case, the hot gases flow through the throat of the rocket nozzle where it expands and develops thrust.

Fig. 1 is a diagram of a simple nuclear rocket system. The nuclear rocket stage consists of a single propellant tank (usually cryogenically stored hydrogen), a nuclear reactor energy source, and turbopump machinery to move the propulsive working fluid into the reactor, where it is then heated and exhausted at very high temperatures through the nozzle assembly. This expansion process of the heated propellant delivers thrust to the rocket vehicle assembly and its payload.

Specific impulse is used as a measure of rocket performance. This is defined as:

$$I_{sp} = \frac{\text{thrust}}{\text{mass flow rate of propellant}} = \frac{F}{\dot{m}}$$

(1)

where F is the thrust in newtons (N) or pounds-force (lb$_f$)
$\dot{m}$ is the mass flow rate of propellant in kilograms per second (kg / s) or pounds-mass per second   (lb$_m$ / s)

In terms of the International System of Units (SI system), the units of specific impulse are:

$$I_{sp} = \frac{F}{\dot{m}} = \frac{\text{newtons}}{\text{kilograms/second}}$$

(2)

NUCLEAR ROCKET
PROPULSION
SYSTEM

PROPELLANT
TANK

PROPELLANT
FEED PUMP

BLEED TURBINE DRIVE

NUCLEAR REACTOR
HEAT EXCHANGER

HEATED PROPELLANT

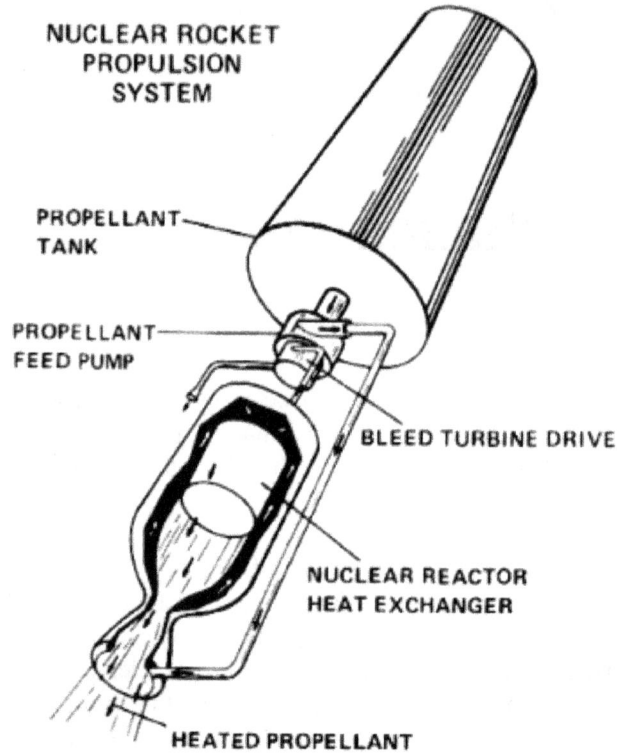

Fig. 1. Typical nuclear rocket propulsion module. *Courtesy of NASA and Department of Energy, After J. M. Taub, 1974*

Since one newton equals one kilogram-meter per second squared (I N = I kg-m / s^2), the basic units for specific impulse then becomes:

$$I_{sp} = \text{meter/second (m/s)}.$$

(3)

In the English system of units:

$$I_{sp} = \frac{\text{thrust}}{\text{mass flow rate}} = \frac{lb_f}{lb_m/s}$$

(4)

Using Newton's second law of motion,

$$F = (1/g_c)\, m\, a$$

$$(1\ lb_f) = (1/g_c)\, (1\ lb_m)\, a$$

(5)

where a   is the acceleration due to gravity (ft / s^2)
       g_c is a universal dimensional constant (32.17 ft-lb$_m$ / lb$_f$-s^2).

At sea level on the surface of the Earth, a = 32.17 ft / s^2 and eq. 5 reduces to 1 lb$_f$ = I Ib$_m$, that is, one pound-force is equal to one pound-mass. Some confusion occurs in interpreting eq. 5, when this condition of equality is forgotten or ignored. Often this relationship between one pound-force and one pound mass is used to simplify the expression found in eq. 4, namely

$$I_{sp} = \frac{lb_f}{lb_m/s} = seconds.$$

(6)

This simplification, strictly speaking, is valid only where 1 lb$_f$ = 1 lb$_m$ that is, the sea level on the surface of the Earth. But, specific impulse is frequently described in terms of "seconds" in the traditional engineering system units.

Performance of a rocket engine is often expressed as

$$I_{sp} = \frac{F}{\dot{m}} = A\, C_f\sqrt{(T_c/M)}$$

(7)

where I_{sp} is the specific impulse (m / s)
       F is the thrust (N)
      $\dot{m}$ is the mass flow rate of propellant (kg / s)
      A is a performance factor related to the thermophysical properties of the propellant
      C_f is the thrust coefficient which is a function of the nozzle parameters
      T_c is the chamber temperature (K)
      M is the molecular weight of the exhaust gases.

Eq. 7 reveals that the specific impulse, and therefore the engine performance, increases with higher chamber temperatures propellant gases and lower molecular weight exhaust gases. As a result, hydrogen (H$_2$), as the lowest molecular weight gas (M = 2), is the working fluid of choice for nuclear rockets. The hydrogen propellant is usually stored as a liquid at cryogenic conditions in order to minimize the volume and consequently the size of the spacecraft stage. Launch vehicles dictate the size of individual propellant storage tanks that can be launched from Earth to orbit. However, several propellant tanks can be launched separately from Earth and then assembled into clusters on orbit. Therefore, nuclear staged propulsion systems can be configured to be of virtually any size.

The basic nuclear rocket engine concept is quite simple. As seen in Fig. 1, it consists of a nuclear reactor used to heat a low-molecular-weight gas (hydrogen) to as high a temperature as possible, a nozzle through which the gas expands, and a turbopump to force the propellant through the system. However, in order to obtain the desired stage performance, the actual components become quite complex. The reactor must operate at very high temperatures for high performance and high power densities to minimize the effect of their weight. This combination of high temperature and high power density is the challenge in the design of thermal nuclear rockets.

During operation of a representative solid-core nuclear rocket (see Fig. 2), the hydrogen propellant is pressurized and fed through the nuclear reactor by a turbopump. This high pressure working fluid protects the nozzle from melting by regeneratively cooling the nozzle. The high pressure fluid is also used to cool the reflector on its way to the reactor. Depending on the nuclear rocket cycle, some hydrogen can be bled from the propellant flow system at a number of points, such as the thrust chamber or reactor outlet region, to run the turbopumps. However, the main quantity of propellant flows through the reactor core, which, as configured here, contains solid fuel elements arranged in a hexagonal matrix. Holes in the fuel elements are used as coolant passages. The hydrogen propellant, heated to temperatures of 2,400 to 3,100 K, is then expanded and exhausted through the nozzle. This is the configuration used in the NERVA engine (to be discussed in Chapter 3). Alternate nuclear rocket configurations and fuel arrangements will be examined in other chapters.

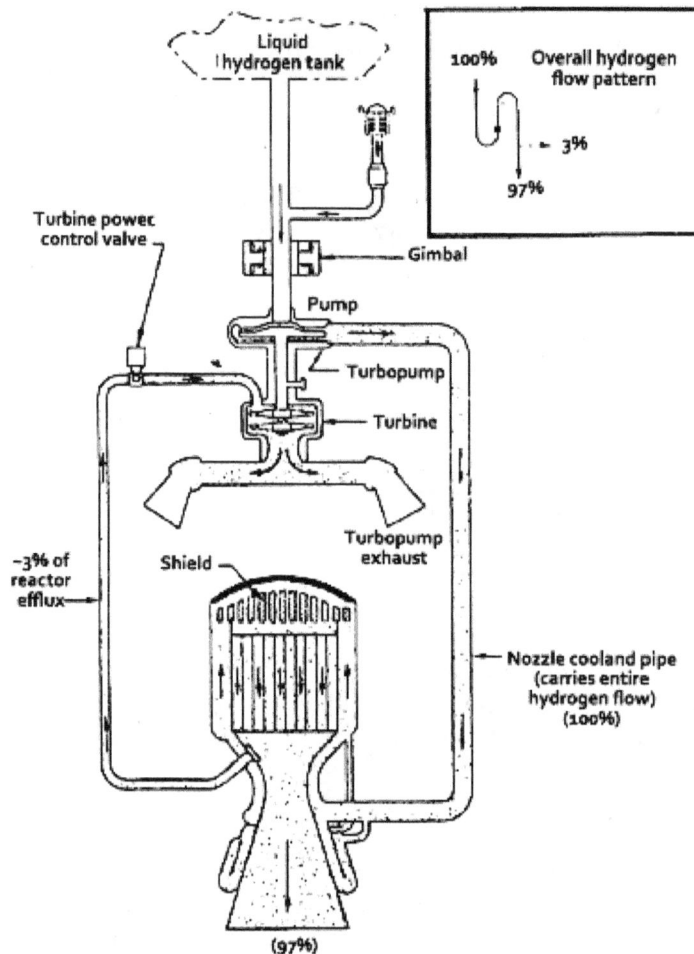

Fig. 2. A "hot bleed cycle" nuclear rocket reactor engine in which a small fraction of the reactor power is used to drive the propellant turbopumps. *Courtesy of Department of Energy.*

The need for very high temperatures limits the reactor designs to a few materials, such as the refractory metals and graphite. The metals are strong neutron absorbers, whereas graphite is not. Graphite, in addition to having excellent high-temperature strength, also acts as a neutron moderator and minimizes the amount of enriched uranium required in the reactor core. However, a great disadvantage of graphite is that it reacts with hot hydrogen to form gaseous hydrocarbons and, unless protected, quickly erodes away the fuel elements. One of the greatest challenges in nuclear rocket programs is the development of fuel elements of adequate lifetime in a high pressure, hot hydrogen environment.[1]

As shown in Fig. 3, solid-core nuclear rockets provide about twice the specific impulse of the best chemical rockets. With the idea of eventually performing manned expeditions to Mars, in 1955 a research program was initiated on the development of hydrogen-cooled, solid core nuclear rockets. Nuclear rocketry is viewed as a necessary part of such interplanetary expeditions. The high specific impulse nuclear propulsion devices greatly reduce the overall transit times to Mars as well as greatly reducing the initial mass that is needed to be delivered to low Earth orbit at the start of the expedition. This, in turn, alleviates the strain on other systems, such as the life support system. For example, the ratio of take-off mass from the Earth's surface to the final mass that achieves Earth escape velocity is a factor of about 15 for a chemical rocket and about 3.2 for a launch vehicle using a solid-core nuclear rocket upper stage. Other forms of advanced propulsion systems are also presented in Fig. 3. Molten and gaseous nuclear reactor cores will be discussed. However, the development of these forms of reactor cores are very much in the research phase. Even more advanced concepts, such as direct fission, thermonuclear fusion, and annihilation of anti-matter, are in a very preliminary conceptual stage and considered beyond the scope of this book.

BOOK 2
NUCLEAR THERMAL PROPULSION SYSTEMS

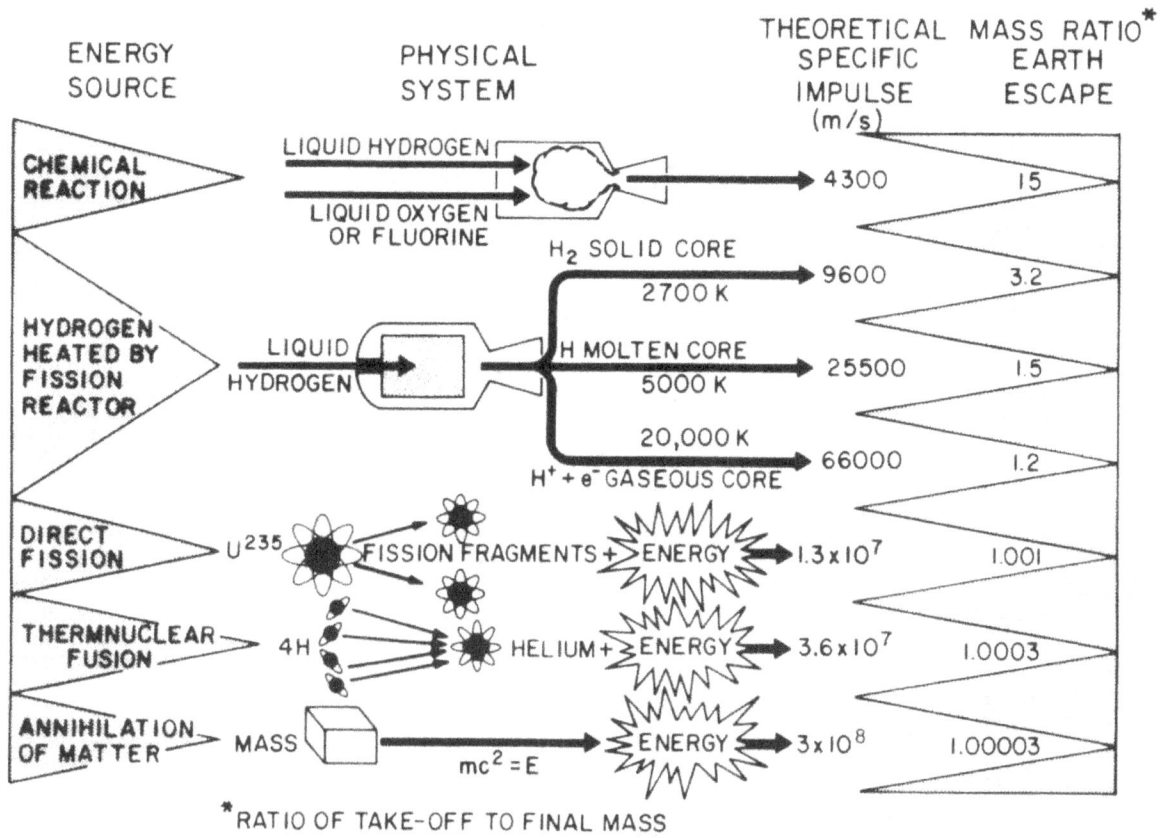

ENERGY SOURCE	PHYSICAL SYSTEM		THEORETICAL SPECIFIC IMPULSE (m/s)	MASS RATIO* EARTH ESCAPE
CHEMICAL REACTION	LIQUID HYDROGEN LIQUID OXYGEN OR FLUORINE		4300	15
HYDROGEN HEATED BY FISSION REACTOR	LIQUID HYDROGEN	H_2 SOLID CORE 2700 K	9600	3.2
		H MOLTEN CORE 5000 K	25500	1.5
		20,000 K $H^+ + e^-$ GASEOUS CORE	66000	1.2
DIRECT FISSION	U^{235} FISSION FRAGMENTS + ENERGY		1.3×10^7	1.001
THERMNUCLEAR FUSION	4H HELIUM + ENERGY		3.6×10^7	1.0003
ANNIHILATION OF MATTER	MASS $mc^2 = E$ ENERGY		3×10^8	1.00003

*RATIO OF TAKE-OFF TO FINAL MASS

Fig. 3. Propulsion performance for various energy sources. *After J. M. Taub,1974.*

Fundamentals of Nuclear Reactors[2]

Reactor Elements

The heart of a fission nuclear rocket is the nuclear reactor. A nuclear reactor is device in which the controlled fissioning or splitting of certain atoms occurs. A product of the fissioning process is the release of significant amounts of energy. A fissile nuclide is a nuclide (e.g., $^{233}_{92}U$, $^{235}_{92}U$. or $^{239}_{94}Pu$) that is capable of being split or fissioned upon absorbing neutrons of any energy. Thus, a nuclear reactor is configured to contain fissile nuclear material, such that a chain reaction of fission events can be maintained and controlled.

A chain reaction is a reaction that stimulates its own repetition. For a fission chain reaction, a fissile nucleus absorbs a neutron, splits, and releases additional neutrons (see Fig. 4). A fission chain reaction is self-sustaining when at least one neutron per fission event survives to create another fission reaction. The two lighter elements produced in the splitting of the heavy nucleus are called fission products (FP).

CHAIN REACTION

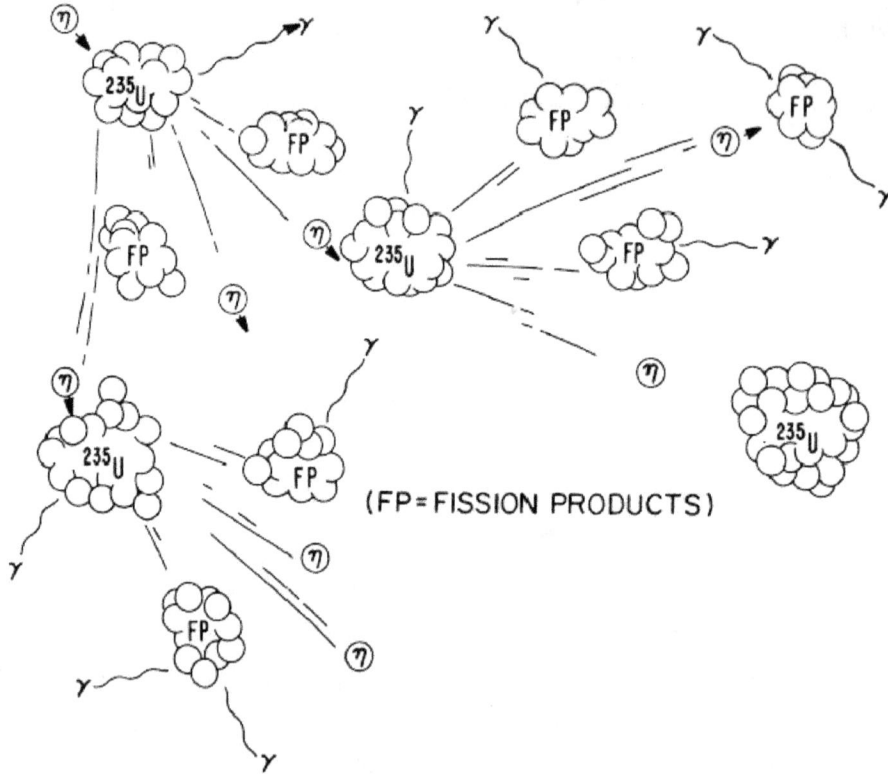

Fig. 4 Fission chain reaction.

The multiplication factor, k, is used to describe the fission chain reaction. The definition of k is: [3, 4, 5]

$$k \equiv \frac{\text{number of nuclear fissions (or neutrons) in one generation}}{\text{number of nuclear fissions (or neutrons) in the immediately preceding generation}} \qquad (8)$$

Figure 5 illustrates the behavior of the fission chain based on the value of the multiplication factor k. When
$\quad k = 1$, the fission reaction is critical or self-sustaining and power production occurs at a steady rate.
$\quad k < 1$, the chain reaction is called subcritical and the number of fissions occurring per generation (or the neutron population) will eventually reach zero.
$\quad k > 1$, the chain reaction is supercritical and the number of fission reactions (or the neutron population) increases each generation (see Fig. 5)

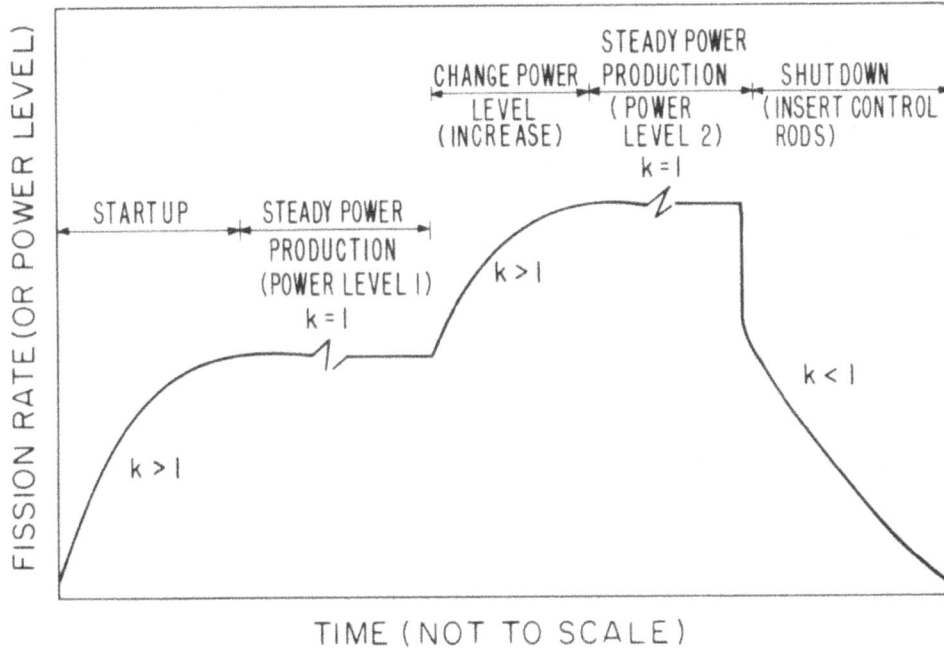

Fig. 5. Multiplication factor as a function of reactor conditions.

Space nuclear reactor components include: a core with the nuclear fuel of fissile material, coolant to remove energy generated in the core, a reflector to minimize the lost of neutrons from the core, control drums or rods to regulate the multiplication factor, and a radiation shield to protect components outside the reactor from destructive radiation.[6, 7, 8, 9] (see Fig. 6). In addition, a moderator component maybe present to regulate the neutron spectrum in the region of nuclear fissioning. The predominant fissioning energy spectrum depends on the reactor design. A thermal reactor is one in which thermal neutrons are the predominant cause of fission reactions, while a fast reactor is one designed so that the majority of fissions occur at fast neutron energies, for instance, > 100 keV. A moderator is found only in thermal reactors.

The arrangement of the nuclear reactor has the core as the central reactor region. It contains the nuclear fuel, a moderator (thermal reactors only), suitable structural materials, and coolant passages for heat removal. The nuclear fuel fissile material in space reactor is usually highly enriched (typically 93.5 percent) uranium-235. It sustains the fission chain reaction, is responsible for the criticality of the reactor, and provides for the release of large quantities of energy in the nuclear fission reaction. Fuel may be in solid, liquid, or gaseous form. If a moderator is present, it is a low-mass material, such as hydrogen, that slows down or moderates neutrons from a fission energy spectrum to a thermal energy spectrum, or perhaps an epithermal spectrum. (Epithermal neutrons are neutrons with energies above thermal values.) This moderating material is found only in thermal or epithermal reactors and is not used in fast reactors.

Coolant is used to remove the thermal energy from the reactor core. This heat removal function is accomplished using a pumped-working fluid, usually hydrogen, in nuclear rockets.

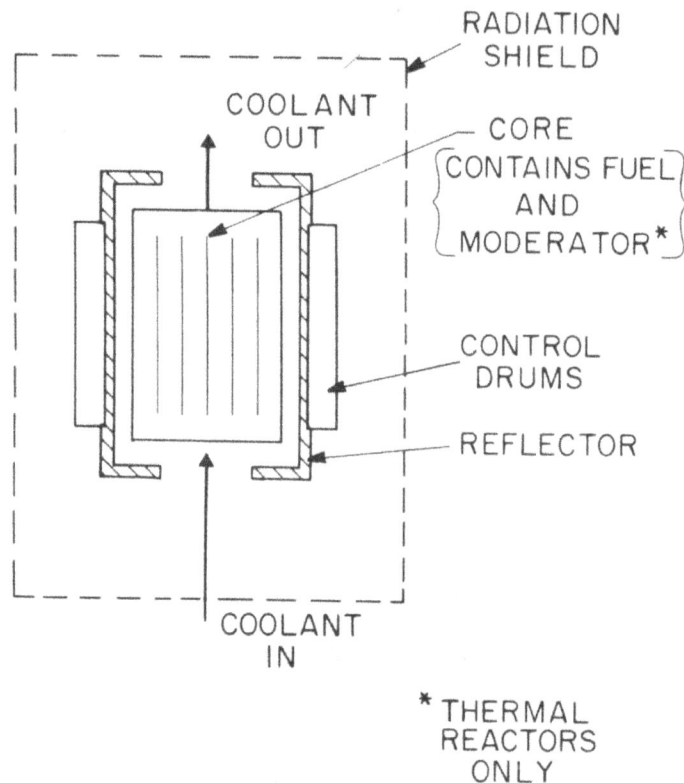

Fig. 6. Generic space nuclear reactor system.

The reflector is a material that scatters neutrons back into the core. It is located adjacent to the reactor core. Beryllium and graphite are typical reflector material. These materials have high scattering cross section and low neutron absorption characteristics. The reflector reduces neutron leakage and permits a more uniform power production within the core through flux flattening. Neutrons leaking out of the core scatter in the reflector material and some of these return to the core. The use of a reflector helps reduce the critical mass of the system and supports flux flattening and a more uniform power generation in the core.

The reactor is controlled by various types of neutron absorber materials. These can be arranged in several forms, such as rods in the core or drums in the reflector. Boron carbide (B_4C) is a representative material used to regulate the nuclear reactor. Movement of these rods or drums adjusts the multiplication factor, thereby controlling the reactor's power level (see Fig. 5).

Astronauts and equipment outside the reactor must be protected from neutrons and gamma rays escaping the reactor core. A protective radiation shield is used for this purpose. In nuclear rockets, only partial shields are needed with the reactor located at the end of the spacecraft. Typical space reactor shield materials include lithium hydride (LiH) for neutron attenuation and tungsten (W) for gamma ray absorption. The hydrogen propellant tank also serves as a shielding device.

The Fission Process

A nucleus must be in an excited state for the nuclear fission process. Fission may occur if the excitation energy exceeds a certain critical energy. Or, emission of gamma radiation to return the compound nucleus to its ground state may occur. Fission threshold energy are shown in Table 1 calculated for a wide range of atomic masses.[10, 11] Only the very heavy nuclides (mass number (A) > 230) have reasonably low threshold energies. Experimental values of the fission thresholds for selected heavy nuclides are presented in Table 2. Negative threshold energy (i.e., $^{233}_{92}U$, $^{235}_{92}U$, or $^{239}_{94}Pu$) mean that neutrons with essentially zero kinetic energy (thermal neutrons) can cause these nuclides to undergo "thermal" fission. $^{233}_{92}U$, $^{235}_{92}U$. and $^{239}_{94}Pu$ are called "fissile" nuclides while the other heavy nuclides shown (i.e., $^{232}_{90}Th$, $^{234}_{92}U$, $^{236}_{92}U$, and $^{237}_{93}Np$) can only undergo fast fission.

Table 1 Neutron fission thresholds as a function of nuclear mass (calculated).

Mass Number (A)	Fission Threshold (MeV)
16	18.5
60	48
100	47
140	62
200	40
236	~ 5

Table 2 Neutron fission thresholds of heavy nuclides (experimental)

Target Nucleus	Compound Nucleus	Fission Threshold (MeV)
^{232}Th	^{233}Th	1.3
^{233}U	^{234}U	< 0
^{234}U	^{235}U	0.4
^{235}U	^{236}U	< 0
^{236}U	^{237}U	0.8
^{238}U	^{239}U	1.2
^{237}Np	^{238}Np	0.4
^{239}Pu	^{240}Pu	< 0

Fission causes the highly unstable compound nucleus to almost always splits into two fission fragments. The nucleons are more tightly bound within the fission fragments than they were in the original heavy nucleus. Large amounts of energy (typically 200 MeV per fission) is released in this nuclear reaction. This energy is the difference in binding energy between the original heavy nucleus and its fission products. Table 3 presents a typical energy distribution for uranium-235 fission. Nuclear fission results in the prompt emission of several neutrons. The average number of neutrons released per fission (v) is a function of the neutron energy and the fissile nuclide. Thermal fission of uranium-235 releases about 2.5 neutrons per fission. In addition, some of the fission products themselves are nuclides which possess more neutrons than necessary for nuclear stability. These unstable fission fragments (with half-lives of up to approximately one minute) emit delayed neutrons. For thermal fission of uranium-235, the delayed neutron fractions (β) are about 0.7 percent (see Table 4).[12] Because the fission fragments are neutron-rich, they undergo radioactive decay, emitting mainly beta and gamma radiations.

Table 3. Typical energy distribution for Uranium-235 fission.

Energy Form	Energy Released (MeV)	Energy Potentially Recoverable (MeV)
Kinetic Energy of Fission Fragments	168	168
Decay of Fission Products		
- Beta Radiation	8	8
- Gamma Radiation	7	7
- Neutrinos	12	--
Prompt (Fission) Gamma Radiation	7	7
Kinetic Energy of Fission Neutrons	5	5
Capture Gamma Radiation	--	3-12
Totals	**207**	**198-207**

Table 4. Delayed neutron fraction (b) for fast and thermal fission.

Nuclide	Fast Fission	Thermal Fission
^{233}U	0.0027 ± 0.0002	0.00264 ± 0.0002
^{235}U	0.0065 ± 0.0003	0.0065 ± 0.0003
^{238}U	0.0157 ± 0.0012	—
^{239}Pu	0.0021 ± 0.0002	0.0021 ± 0.0002
^{240}Pu	0.0026 ± 0.0003	—

Fig. 7. illustrates the fission product mass numbers as a function of fission yield for the thermal and fast (14 MeV) fission of uranium-235. The fission yield is defined as the percent of the total number of fissions that produces fission products of a given mass number. As the energy of the incident neutron increases, the probability of producing two fission products of nearly the same size increases and the characteristic dip that is present in the thermal fission-yield plots at approximately mass number 120 disappears.

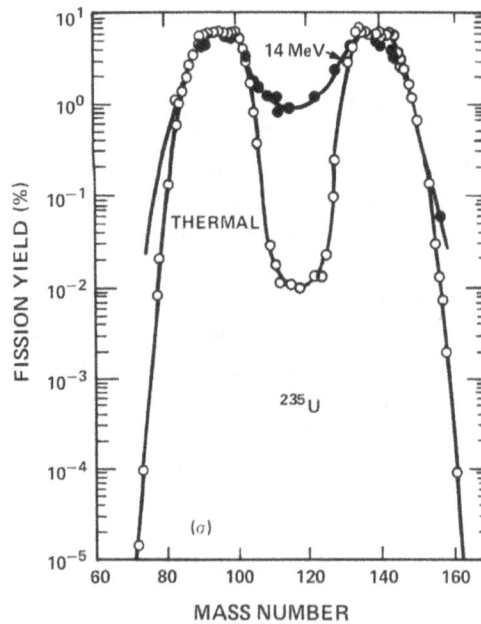

Fig. 7. Fission product mass distribution for uranium-235 fission.

BOOK 2
NUCLEAR THERMAL PROPULSION SYSTEMS

Afterheat

Afterheat or decay heat is the thermal energy released by the continuing radioactive decay of nuclides (mostly fission products) in a reactor core after the fission chain reaction has been terminated. After shutting down an operated reactor by making the fission process subcritical, the reactor will continue to generate thermal power for a period of time. This results from the decay of the accumulated fission products, primarily gamma and beta emitters. Post-operational thermal energy release is a function of the power level at which the reactor operated before shutdown and the length of time it operated at this power level. These parameters define the fission product inventory. Fig.8 illustrates the decay energy from a nuclear rocket (NERVA engine) after a one-hour full power operating time.

As a consequent of the decay heat, a nuclear rocket does not have a sharp shutdown function like a chemical rocket. In a chemical rocket, the combustion process can be sharply terminated. In a nuclear reactor, coolant must be provided to the nuclear reactor to remove the afterheat in order to avoid reactor   destruction. This coolant can take the form of added propellant that can be account for in the spacecraft velocity computations or a secondary coolant loop to generate electric power to the spacecraft.

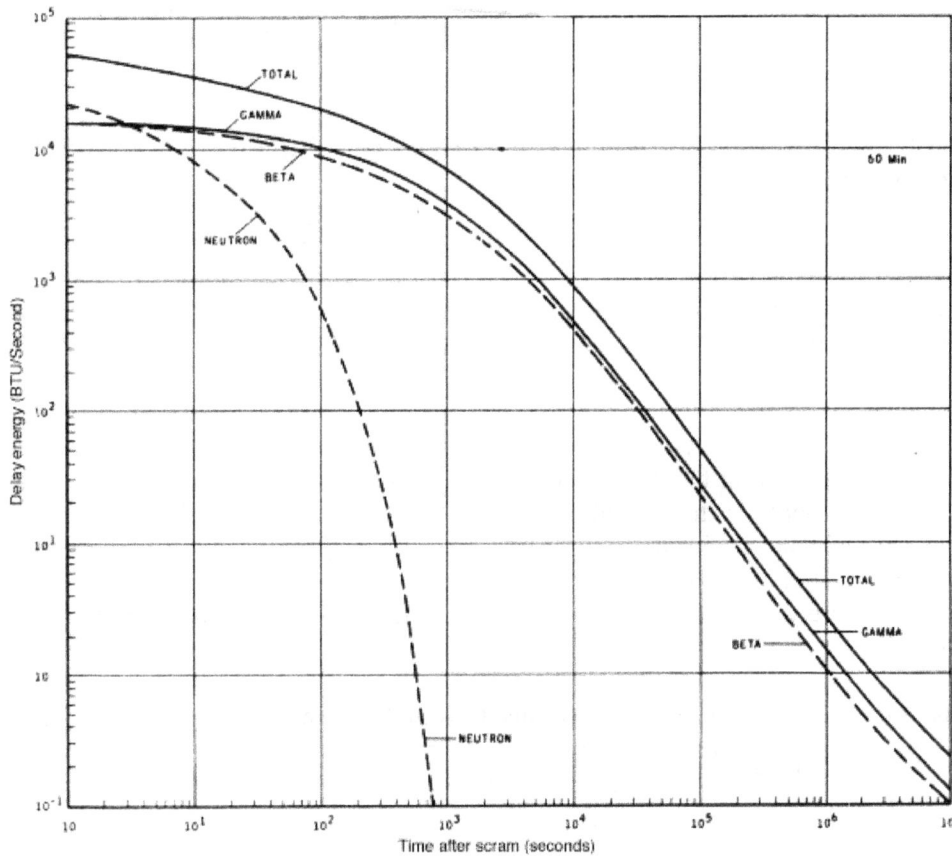

Fig. 8. Decay heat for 60 minute operating time of nuclear rocket.

Shielding

Radiation shielding is used to protect the astronaut crew, payload and radiation-sensitive spacecraft equipment from the damaging effects of radiation. Fig. 9[13] presents typical radiation damage thresholds for certain materials, components, and systems that might be used in a nuclear-powered spacecraft. When such materials and components are exposed to nuclear radiation, their properties are changed in a variety of ways including rate effects and cumulative effects that reduce their overall established lifetimes.

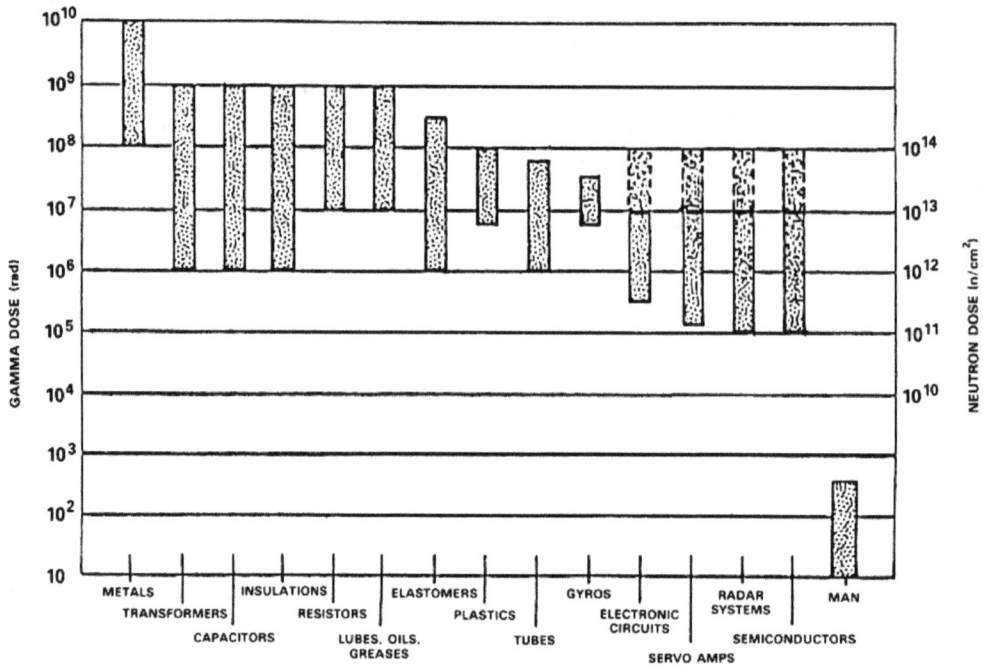

Fig. 9. Typical radiation damage thresholds. *Courtesy of U. S. Department of Energy and Rockwell International.*

The rate effect usually involves the intense radiation dose in a short period of time that leads to ionization-induced damage. For example, sensitive electronic circuits will experience noise, electronic up-set, or even burnout under influence of a rate effect. Cumulative radiation typically involves the long-term buildup of radiation-induced lattice defects in the material. This produces significant changes in thermophysical properties.

As Fig. 9 illustrates, humans are more sensitive to radiation damage than material components. Table 5 list some of the effects of radiation on the human body. As a point of reference, the estimated annual average wholebody dose rates to people in the United States is ~ 180 mrem / yr or ~ 1.8 mSv / yr. To reduce the radiation levels to acceptable levels for the crew, radiation shielding is included in the spacecraft. The absence of a surrounding gaseous medium or atmosphere which would scatter neutrons back toward the system makes it feasible to design the crew radiation shield within a protected cone located between the reactor and other elements of the spacecraft.

Table 5. Effects to the body from localized exposure to X and gamma radiation.

Organ of Body	Acute Irradiation Level (rem)	(sievert)	Acute and Delayed Biological Effects
SKIN	300	3.0	Erythema or "sunburn" effect noticeable.
	1500	15.0	Raw, moist skin surface where irradiated.
	5000–7000	50–70	Ulceration, slow healing, possible skin cancer
GONADS	50	0.5	Brief functional sterility in males only.
	250	2.5	Sterility for 1 to 2 years in both male and female.
	600	6.0	Permanent sterility.
EYE	200	2.0	Change in optic lens opacity.
	600	6.0	Clinically significant cataract.
FETUS	10–20	0.1–0.2	Significant probability of malformation, if irradiation occurs in first 3 months of pregnancy.

BOOK 2
NUCLEAR THERMAL PROPULSION SYSTEMS

Shielding materials are often chosen on the basis of the following factors:

1. Ability to attenuate radiation
2. Minimum mass
3. Resistance to radiation-induced thermophysical damage
4. Stability at elevated operating temperatures, i.e., 700-900 K regime
5. Ease of fabrication
6. Availability
7. Cost

Materials containing large amounts of hydrogen make the best neutron attenuating materials. One of the most important parameters for rating shielding materials for space applications is the number of hydrogen atoms (N_H) the material contains per unit volume. Water, with an N_H of approximately 6.7×10^{22} atoms of hydrogen / cm^3, is commonly used to moderate and shield low temperature terrestrial reactors, but it is not suitable as a shield above its critical point. Because of their high temperature stabilities and hydrogen content values, metal hydrides are the prime candidates for neutron shielding applications in space. Of these materials, lithium hydride (LiH) is the favored choice because of its high hydrogen density ($N_H = 5.9 \times 10^{22}$ hydrogen atoms / cm^3), low mass density (0.775 g / cm^3), and moderately high melting point of ~ 960 K. Lithium hydride also produces a minimum amount of secondary radiation. Some LiH thermophysical properties are presented in Table 6, while Fig. 11 depicts typical LiH neutron attenuation. The thermal environment of a space reactor shield is a critical design consideration. For example, when LiH melts, it expands some 25 percent. Consequently, the shield must be kept below the LiH melting point temperature to avoid undesirable thermomechanical stresses.

Table 6. Representative properties of lithium hydride (LiH).

Density (g / cm^3)	0.775
Molecular weight	5.95
Hydrogen content N_H (atoms H / cm^3)	5.85×10^{22}
Hydrogen content (weight %)	12.68
Melting point (K)	960
Volume change on melting (%)	$+25 \pm 2$
Crystal Structure	Face centered cube (fcc)
Heat of fusion (kJ / mole)	21.77 ± 1.3
Compressive strength, cold pressed (MPa)	96.5 - 165.5
Molar volume (cm^3 / mole)	10.254

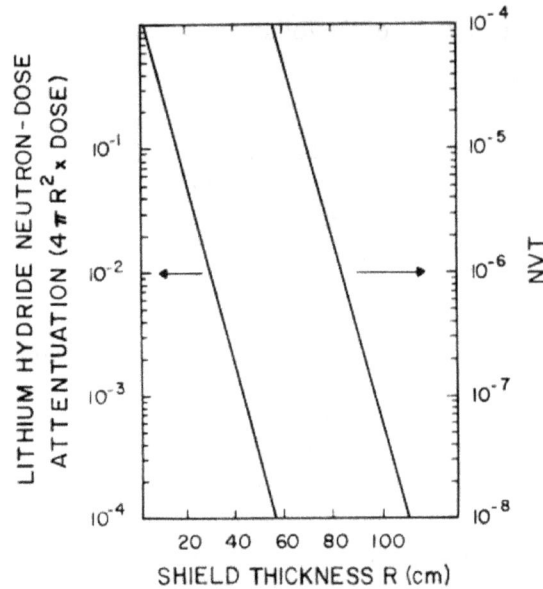

Fig. 10. Typical lithium hydride neutron attenuation. *Courtesy of Los Alamos National Laboratory.*

If natural lithium is used as the shield material, tight containment of the lithium hydride is a shield design complication in high neutron flux regions. In natural lithium, approximately 7.5 percent of the lithium has a very high thermal neutron cross section-that is, 945 barns at 2200 m / s for the ^{6}Li$(n,\alpha)^3$H reaction.

$$^6_3\text{Li} + ^1_0\text{n} \rightarrow ^3_1\text{H} + ^4_2\text{He}.$$

(8)

The helium generated by this reaction would build up significant pressures within the LiH shield containers during the lifetime of the space power system. Consequently, to avoid this problem, LiH isotopically enriched in the $^7_3$Li can be used in shield regions where the thermal neutron fluence is expected to be high. The $^7_3$Li is generally enriched to 99.99 weight percent.

Lithium hydride must be loaded, processed, and when at room temperature, maintained in a very dry atmosphere to prevent reaction with water or water vapor. At elevated temperature, a hydrogen overpressure must be maintained to prevent hydrogen loss.

If shielding is required to protect against gamma radiation, high density materials are used. These minimize overall shield volume and system mass. Table 7 lists properties for some of the high density, high melting point materials considered as suitable for gamma ray shielding materials.

Table 7 Thermophysical properties of candidate gamma ray shielding materials.

Property	Uranium-8 wt% Molybdenum (U-8Mo)	Tantalum-10 wt% Tungsten (Ta-10W)
Density (g/cm^3) (at room temperature)	17.40	16.83
Thermal expansion (10^{-6} cm/cm-K)	14.4 at 673 K	6.3 at 373 K / 6.9 at 773 K
Thermal conductivity (J/s-cm-K)	1.34 at 298 K / 2.89 at 773 K	0.762 at 298 K / 0.695 at 773 K
Specific heat (kJ/kg-K)	0.138 at 298 K / 0.176 at 773 K	0.150 at 298 K / 0.147 at 773 K
Melting point (K)	1403	—

In summary, shield design for space nuclear rockets depends to a large extent on whether the mission will be manned or unmanned. Design alternatives are also driven by the desire to minimize mass, using a shadow shield configuration. Since the reactor system is generally smaller in size than the payload/spacecraft, the shield mass becomes a direct function of the solid angle subtended by the shadow shield. Consequently, minimum mass shadow shields require that the space reactor itself possess a minimum projected area and that all structural components of the spacecraft be contained within the conical region established by the dose plane and the extremities of the radiation source. The hydrogen propellant tank will probably be truncated at the end near the reactor to minimize the shielding mass.

Shield design for reactors used in manned missions is driven by much lower permissible radiation dose levels and the reactor shield must include both neutron and gamma ray attenuation components. In general, this type shield will be a laminated one with layers of heavy metal gamma ray shielding followed by tightly packaged containers of neutron attenuating lithium hydride. Secondary gamma rays from neutron radiative capture reactions can also make a significant contribution to the radiation level appearing in the dose plane and must be compensated for.

Another factor to be considered in protecting the crew from the reactor radiation in a nuclear thermal rocket is that there will probably be a need for a "storm cellar" to protect the crew from solar flares. During the short nuclear thermal reactor operational times, the crew could use this "storm cellar" to shield them from the reactor radiation in total or in part to minimize the need for additional shielding.

Chapter 2

Nuclear Thermal Propulsion Requirements

The particular mission applications will establish such performance parameters as the specific impulse, mass, and size of the nuclear rocket requirements. Safety will add additional requirements to achieve required operational needs and will probably specify the need for certain redundant elements. Reliability requirements will result in the need for a complete understanding of all component behavior, element interactions, development of much more robust designs, and may add the need for redundancy into the engine or spacecraft design. Cost, schedule and space qualification experience for manned missions all are factors in the establishment of nuclear thermal propulsion requirements. When feasible, development of nuclear rockets should consider being able to perform a wide-range of missions.

Mission Applications

Interest in nuclear thermal rockets has centered on manned trips to Mars. For this application, nuclear rockets sufficiently reduces the amount of mass in Earth orbit at the beginning of the trip to call it an "enabling" technology--a technology that one must have. The original nuclear rocket development program, under Project Rover/Nuclear Engine for Rocket Vehicle Applications (NERVA) was planned to operate with Saturn V rockets and assemble a Mars vehicle in low Earth orbit. With the termination of the Saturn V rocket program, the manned Mars exploration program of the 1970s was also ended. Project Rover/NERVA subsequently was terminated in 1973.

Renewed interest in Mars occurred when in 1989, President George Bush challenged America in a way no one has challenged us before: "Back to the Moon to stay, and onward to Mars." This Space Exploration Initiative established a group under Gen. Thomas P. Stafford to evaluate technology options to accomplish the Mars goals. They concluded that, "As result of the challenges of a Mars trip, several hundred tons of equipment and fuel are required for a Mars expedition. Thus, we will require a conventionally powered heavy lift launch capability to minimize assembly in Earth orbit. From Earth orbit to Mars, nuclear propulsion technology will allow reduced weight, approximately one-half that of chemical systems, and achieve faster interplanetary trip times."[1] A typical mission profile is shown in Fig. 1.

Other missions can take advantage of nuclear rockets developed for manned Mars exploration. For instance, trips to the Moon can use the higher performance of nuclear rockets to transport large habitat modules or science laboratories. This is a good way to obtain the operating experience prior to commitment to a multi-year mission to Mars. Much more robust planetary science missions also become feasible (See Fig. 2).[2]

In the 1987-1992, the Strategic Defense Initiative did some development work on a fast response missile interceptor with a nuclear thermal rocket stage. The potential high performance of the nuclear rocket concept made this an interesting idea to pursue at that time.

In January 2004, President George W. Bush stated renewed interest in returning to the Moon and preparing for "journeys to the worlds beyond our own." As such, it is expected at some future time, there will be renewed programs in space nuclear rockets.

Using Mars as the eventual goal and the desire to gain operational experience prior to such a mission, questions to be address include what rocket engine, specific impulse, engine thrust level, and operating times will be of primary interest. Many parametric studies have been performed to address these issues.[3,4,5]

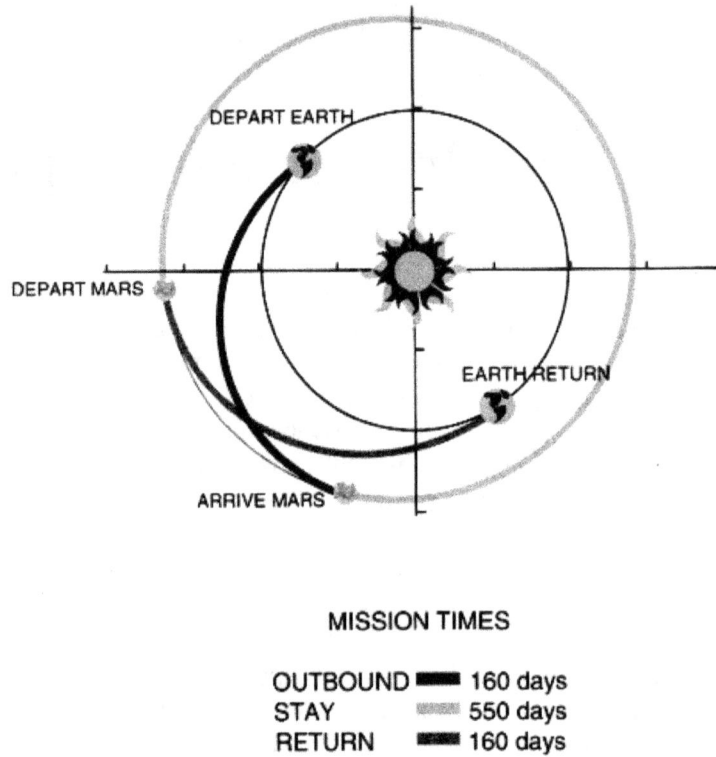

MISSION TIMES

OUTBOUND ▬▬ 160 days
STAY   ▬▬ 550 days
RETURN   ▬▬ 160 days

Fig. 1. Typical nuclear propulsion mission to Mars. Total mission:870 days, 2014 shortened transit time. *America At The Threshold*

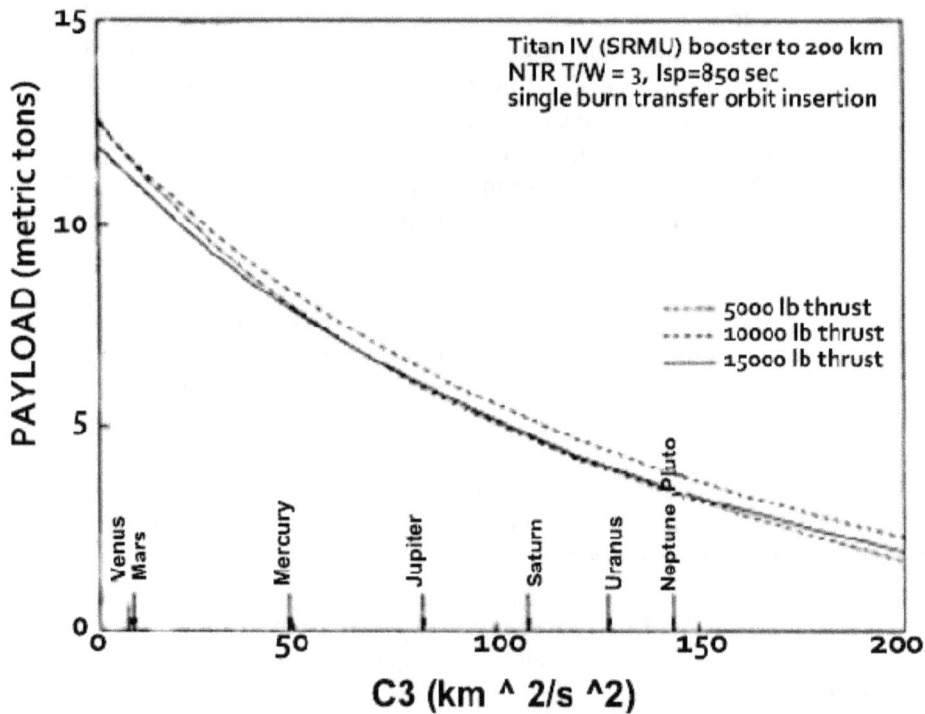

Fig. 2. Small nuclear thermal rocket interplanetary performance. *From David Buden, 1992.*

Higher specific impulse results in decreased amounts of propellant needed for a given mission. Fuel materials limit the propellant temperatures to the 2,450 to 3,200 K range, with corresponding specific impulses between 800 and 1000 s. (See Fig. 3).

Thrust levels are trade-offs with engine burn times and engine weight. Original Mars trips considered conventional single burn trajectories vehicles. A vehicle gross weight of 227,000 kg (500,000 pounds) required a thrust of about 114,000 kg (250,000 pounds) with a reactor power level of 5,000 megawatts. This size reactor was actually tested as part of the Rover program. However, the understanding of perigee propulsion (see Fig. 4) reduced the required thrust level without performance penalty to 65.8 kN (15,000 pounds) for a reactor power of 300 megawatts.[6] Considering burn times, thrust levels of interest are usually in the 65.8 to 439 kN (15,000 lb_f to 100,000 lb_f) range. Full power operational times for manned Mars missions are approximately one to five hours. As part of the trade-offs, engine weight must be considered in any optimization because it increases as a function of thrust level.

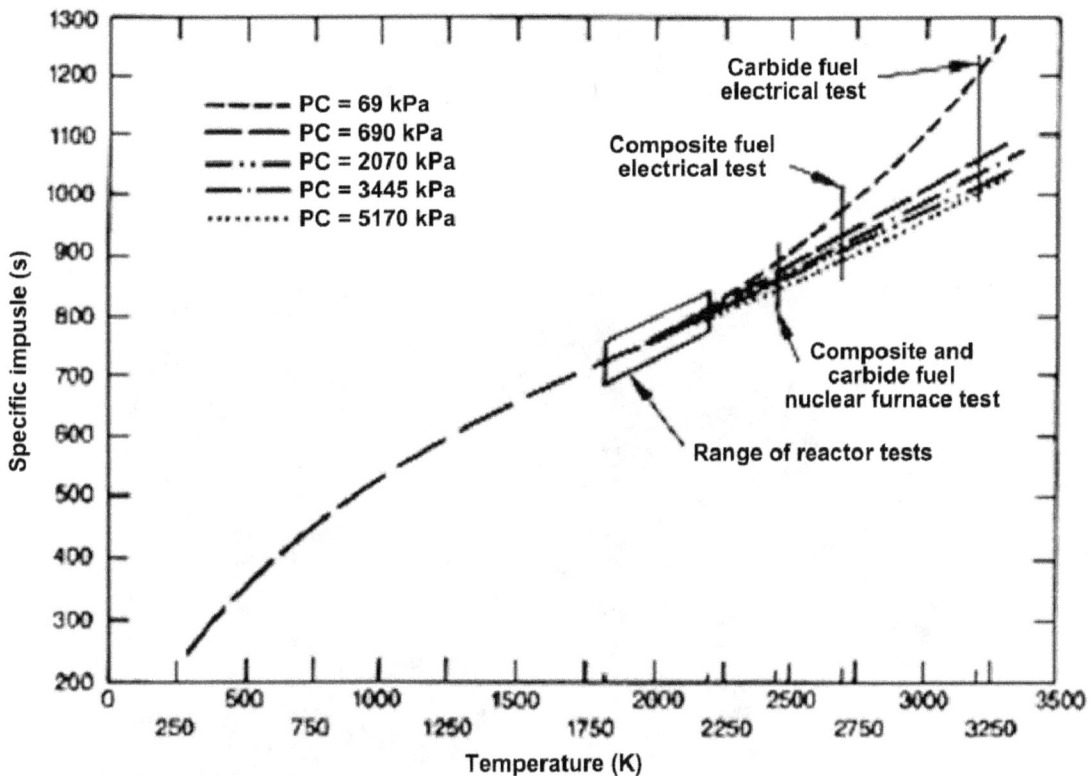

Fig. 3. Specific impulse versus exhaust nozzle temperature for nozzle area ratio 300. *From Jack Ramsthaler and David a. Baker, 1989.*

There were three significant nuclear rocket engine development programs--two by the U.S.A. and one by Russia. From 1955-1973, the Rover/NERVA program developed technology for manned missions to the planet Mars. Also starting in 1955 and continuing through the fall of the U.S.S.R., Russia was active in developing nuclear rockets. In addition to these programs, the U.S.A. in the 1987-1991 time period performed developmental work on a possible anti-missile interceptor using a nuclear thermal rocket stage. The following chapters will cover these developmental efforts in some detail. Table 1 summarizes some key parameters for these activities.

The Rover/NERVA program demonstrated a full experimental nuclear rocket as part of the program. Fig. 5 shows the experimental rocket engine in the test facility at the Nuclear Rocket Development Station in Nevada. The Russians demonstrated all of the key components and an electrically heated test bed engine in their program. Fig. 6 shows the Russian RD-0410 NTP engine hardware. The anti-missile technology program was still in the fuel element development stage at cancellation.

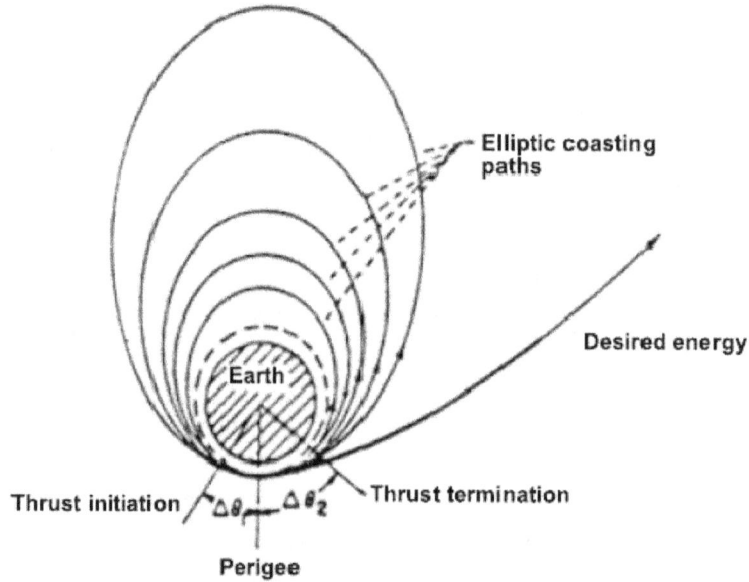

Fig. 4. Sample perigee propulsion trajectory. *From Frank E. Rom, 1991.*

Fig. 5. NERVA *experimental* **engine XE installed in Engine Test Stand No. 1 at the Nuclear Rocket Development Station in Nevada.** *Courtesy of NASA*

Fig. 6. Russian RD-0410 NTP engine. *Credit: Dietrich Haeseler*

Table 1. Nuclear thermal rocket engine parameters for major programs.

	NERVA	Small NERVA Engine	RD-0410	Timberwind	Timberwind 45
Country	U.S.A	U.S.A.	Russia	U.S.A.	U.S.A.
Fuel composition	(UC-ZrC)C composite	(UC-ZrC)C composite	(U,Nb,Zr)C	UC-TaC core with layers of PyC/TaC and ZrC	UC-TaC core with layers of PyC/TaC and ZrC
Fuel form	Prismatic	Prismatic	Twisted ribbon	Particle bed	Particle bed
Propellant	LH_2	LH_2	LH_2 + Hexame	LH_2	LH_2
Vacuum thrust [kN]	337	72	35.3	178	441.3
[lb_f]	75.1	16.4	7.9	39.7	98.3
Specific impulse (s)	825	875	910	869	1000
(m/s)	8085	8575	8918	8516	9800
Nozzle expansion ratio	100	100			

	NERVA	Small NERVA Engine	RD-0410	Timberwind	Timberwind 45
Chamber pressure (MPa)	3.1	3.1	7		
Core outlet temperature (K)	2360	2695	3000	2750	
Power density [MWt/liter]		Based on Pewee results, 2.3 avg., 5.2 max.	35-40	40	40
Burn time [s]	36,000 with up to 60 cycles		3600	100	449
Engine diameter [m]			1.2		4.25
Engine length [m]		3.125 folded nozzle, 4.415 skirt in place	3.7		
Mass with shield [kg]	12300	2550	2000		1500
Thermal power [MWt]	1556	365	196		
Development status	Experimental engine tested	Flight engine design uses NERVA technology	Demonstrated fuel and component technologies	Development effort from 1987-1991, frit design still under development	Development ended in 1992, frit design still under development

Safety Issues Relevant To Nuclear Thermal Propulsion[7]

Nuclear thermal propulsion (NTP) is crucial for successful human Mars exploration. The crew of manned space missions must be protected from any harmful radiation. The safety of the crew is greatly enhanced by shorter trip times. This has the effect of reducing the crew exposure to high levels of galactic radiation, reducing the time that solar flares will be a problem, lowering psychological stresses of long periods in confined environments and reducing the time the crew is subjected to possible equipment malfunctions. The nuclear thermal propulsion rocket engine has many fewer moving parts then chemical rockets which it replaces and could, therefore, be more reliable. There is no need for a chemical rocket oxidizer system. Launch windows for departing Earth and for returning from Mars are significantly wider. Also, there are more opportunities to go to Mars. This provides schedule flexibility and reduced the need for potentially hazardous decisions to meet limited Mars opportunities. In addition, with nuclear thermal rockets having two to three times better performance than chemical rockets, less or no assembly is needed in Earth orbit.

This makes the spacecraft more reliable, less costly and easier to meet schedule. In fact, the mass in low Earth orbit will be one-third to one-half of a chemical rocket mission configuration!

Safety is the prime design requirement for nuclear thermal propulsion (NTP). It must be built in at the initiation of the design process. An understanding of safety concerns is fundamental to the development of nuclear rockets for manned missions to Mars and many other applications that will be enabled or greatly enhanced by the use of nuclear thermal propulsion. In the following discussion, a range of topics including safety requirements and approaches to meet these requirements, risk and safety analysis methodology, and life cycle risk assessments will be covered.

Recognizing that safety is the prime design consideration for nuclear thermal propulsion, it is essential to understand what is meant by this. First of all, the Earth's population and environment must be protected from harmful radioactive materials and radiation. In addition to not significantly effecting the safety of the Earth's population and environment, space nuclear propulsion systems should not significantly adversely affect non-terrestrial environments. This includes space and the environments of other celestial bodies such as the moon and Mars. Cradle to grave protection must be included.

Safety Requirements and Approaches

Safety goals and approaches to achieve safety are given in Table 2. These, safety requirements can be summarized as:
- prevent unplanned nuclear reactor criticality; \maintain thrust as needed to assure safe return of crew;
- maintain core integrity (except possibly on planned dispersal on atmospheric re-entry);
- provide for radiological safety in case of random impact location from a launch abort;
- provide for safe reactor disposal;
- prevent premature reentry of the reactor into the biosphere after operation;
- reduce radioactive levels of the exhaust plume to acceptable levels for the environment and spacecraft;
- protect the crew against unacceptable radiation levels;
- provide independent, high reliable, and redundant operational and safety systems; and
- meet spacecraft safety goals such as in case of certain number of failures the mission can be completed and after that the crew is protected.

Table 3 summarizes the crew dose standards and potential hazards from natural radiation. Exposure to solar flares exposes the crew to the highest potential radiation hazard levels; however, these are anomalous events, and crews can be protected by use of shielded storm cellars. The major radiation risk to the crews is from galactic radiation. These exposures are continuous and, due to the high energy levels, are very difficult to shield against. The best solution is to shorten trip times to Mars.

Table 2. Safety Goals and Approaches

Goals	Reasons	Design Approach
Radiation levels sufficiently low prior to launch to avoid special handling precautions	• Protect workers and astronauts	• Not operate reactor (except for zero power testing) until a stable orbit or flight path is achieved • Independent systems to reduce reactivity to subcritical state • Unirradiated fuel that poses no significant environmental hazard
Prevent inadvertent criticality	• Ensure public not exposed to levels of radiation that exceed standards • Protect crews	• Subcritical if immersed in water or other fluid • Significant negative temperature coefficient • Subcritical on Earth impact accident • Independent reactor safety systems • Quality assurance standards • Positive coded telemetry for reactor startup • Redundant control and safety systems • Independent sources of electric power for reactor control, protection, and communication

		systems • Continuous status monitoring
Avoid unplanned core destructions	• Protect space investments and avoid contamination of space environment • Protect crews	•Independent shutdown systems •Independent decay heat removal •Fault detention for reactor •Positive coded signal to operator
		•
Goals	**Reasons**	**Design Approach**
Avoid release of radioactivity by-products in concentrations that exceed radiological stands	Ensure public not exposed to radiation levels that exceed standards and protect biosphere against concentration of radioactive elements above safety standards	• Design fuel elements to meet standards • Orbital boost system for short-live orbits • Design reactor for dispersal or intact reentry if boosters fail
Avoid contamination of biosphere	Ensure public not exposed to radiation levels that exceed standards and protect biosphere against concentration of radioactive elements above safety standards	• Engine command destruct system • Disposal in deep space

Table 3. Radiation limits (based on NCRP #98) and sources from environmental radiation (referenced to Blood Forming Organs (BFO)

TIME PERIOD	**BLOOD FORMING ORGANS (BFO)**
	REM
30 Day	25
Annual	50
Career	100 - 400*

Based on 3% lifetime cancer mortality risk, age and gender dependent, BFO @ 5 cm depth

RADIATION SOURCES

SOURCE	**BLOOD FORMING ORGANS (BFO)**
Galactic Cosmic Radiation	
Solar Minimum	60 REM/year
Solar Maximum	22 REM/year
Solar Flares	
Ordinary Event	13 REM
Anomalous Event	431 REM

*Assumes 3 g / cm^2 aluminum shielding and BFO @ 5 cm depth

Risk and Safety Analysis Methodology

Safety analysis can use deductive or inductive methods. The former looks at specific information and draws general conclusions. An example is Fault Tree Analysis, where one assumes the system being analyzed is in a failed state and determines how it can occur. Inductive logic examines what happens "if"; it evaluates many cases where components are assumed to have failed and then draws conclusions as to the effects. Failure Mode Analysis is an example--it assumes components in a failed state and determines what happens to the system. Approaches to failure analysis used in safety evaluations are summarized in Table 4.

Failure Mode Analysis (FMA) is used to illustrate the steps in a typical safety analysis such as was used in the NERVA program.

1. Obtain the functional and physical description of the design to be analyzed.
2. Define the functional and physical boundaries, i.e., those items to include in this FMA as opposed to those which must be evaluated by other component or system FMA's.
3. Obtain or define probabilistically the input and output requirements.
4. List the Component Mode of Failure, the operating conditions, the condition of success, the general design analysis that will be required, and the reliability allocated to this failure mode.
5. List the component Mechanism (s) of Failure stemming from the success-failure condition, causes, and interactions; give the probability equation; perform the probability analysis (or assess by an acceptable method such as analytical estimation, direct measurement, historical data, or engineering judgment); show the principal distributions; and report the assessed reliability.
6. Determine how the mechanisms relate to one another (e.g., dependently or independently); and combine the individual assessments to find the probability of success (reliability) under the failure mode.

Table 4. Approaches to Failure Analysis

Type	Purpose	Methodology
Preliminary Hazards Analysis	• Initial assessment of potential hazards during early design phases	• Identify hazards • Determine consequences • Classify effects • Evaluate appropriate corrective actions
Fault Tree Analysis	• Top down approach • Evaluate detailed designs and integrate with mission	• Determine events that can lead to failure or accidents • Construct path from basic causes • Determine failure probabilities for causes • Compute probability of system failure or accident
Failure Modes and Effects Analysis	• Bottom up approach • Evaluate detailed designs and integrate with missions	• Establish failure probabilities • Calculate probability of consequences occurring
Event Tree Analysis	• Multi-event, bottom up analysis • Evaluate detailed designs and integrate with missions	• Identify initiating events • Perform failure modes and effects analysis on consequences • Integrate effects into tree • Computer system success and failures

Life Cycle Risk Assessment[8]

Life cycle risk can be thought of in terms of fabrication, transportation to launch pad, pre-launch, launch, operations, stand-by, and disposal. Emphasis here is placed on launch pad, launch and operations, since fabrication and transport are routinely performed on terrestrial reactors. Accident environments result from launch pad explosions or fires, loss of control, land or water impact, random reentry, etc. A series of questions have been formulated to cover different situations, requirements developed, and design options evaluated to see if these can be safely handled. The questions are given in Table 5. For the postulated accident conditions, the primary safety requirements are determined, design options examined, and the experience base reviewed. The results are given in terms of top level summary discussions. Once a particular design is selected for either unmanned scientific or exploration missions or for crew missions to Mars, detailed design and operational solutions will be needed. The important element here is to have examined the key questions in significant depth to show that solutions exist.

Table 5. Safety questions relevant to space nuclear propulsions.

Ground Operations

- What must be done to safely ground test nuclear rockets?
- What special precautions will be needed at the launch pad?
- How will radioactive material contamination at the launch site be avoided in rocket launch pad accidents?
    - Nuclear criticality
    - Fires
    - Explosions
- How will ground testing be handled so that there are not significant additions to the nuclear waste problem?
- Who approves the launch of vehicles with nuclear rockets on-board?

Launch and Space Operations

- How safe is the crew from reactor radiation?
- How safe are flight operations?
- How will inadvertent criticality be prevented and the population/environment protected for launch/ascent accidents?
- If radioactive materials impact on land, what plans exist to clean up contaminated land areas?
- If a reactor is started below a "Nuclear Safe Orbit" (NSO) or "Sufficient High Orbit" (SHO), how can re-entry of a radioactive core be averted?
- Will an operating nuclear rocket affect other satellites and experiments?
- Will nuclear engines release radioactive materials which contaminate near-Earth space?
- What are the plans for final disposal of nuclear engines in space?
- Returning from Mars, how will a nuclear rocket be prevented from impacting the Earth?

Safe Ground Testing of Nuclear Rockets

Safety is the prime requirement in all testing and operational procedures. The established standards for radiation levels and radioactive releases levels must be met. Environmental Impact Statements are needed before testing facilities can be constructed.

To meet future environmental safety standards, radioactive material removal scrubbers will probably be needed to remove fission gases from the engine hydrogen exhaust and to trap any radioactive material released. This was not the case for most of the Rover and NERVA programs nuclear test articles. In those tests the hydrogen propellant was exhausted directly into the atmosphere.

As the hydrogen coolant flows through the reactor core, a certain amount of contamination with fission products and/or fuel particulate is possible. Several approaches to fission-product retention and waste handling are possible including: (1) closed cycles that involve venting the exhaust to a closed volume or recirculating the hydrogen in a closed loop; or (2) open cycle where in real time the effluent is processed and vented.

The effluent treatment system of this exhaust involves four basic functions (see Fig. 7):[9]

- Initial cooling of the hot reactor exhaust temperatures (range of 2,400 to 3,400 K) to temperatures compatible with normal engineering materials. Also, any debris and large particulate ejected from the core must be retained and maintained in a subcritical configuration.

- Intermediate cooling to temperatures at or below atmospheric. This is considered a desirable part of the design process.

- Fission-product retention to prevent uncontrolled release of contaminants to the environment. This stage retains small particulates, halogens, noble gases, and other volatile species.

- Waste stream processing to properly handle retained fission products, cleaned or process hydrogen effluent, and other potentially contaminated fluids introduced in or generated by the system.

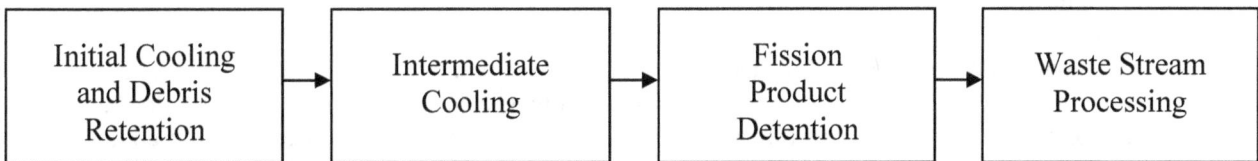

Initial Cooling and Debris Retention	→	Intermediate Cooling	→	Fission Product Detention	→	Waste Stream Processing

Fig. 7. Reactor and engine testing effluent treatment functions. *From* **Larry R. Shipers and John E. Brockmann, 1993.**

The basic hydrogen scrubber technology demonstrated in the Nuclear Furnace-1 testing in 1972 is illustrated in Fig. 8.

Concern also exists in what would happen in a worst case accident. In 1965, in the Kiwi-TNT (Transient Nuclear Test) test, a scenario worse than what is considered the worst credible case, a reactor was intentionally tested to destruction.[10] The primary objective of this experiment was to supply experimental information on the total energy produced, the kinetic or explosive energy released, and the fission product dispersal from a maximum type of accidental reactor excursion. Also, the experiment provided core fragmentation information of interest to the Self Destruct Concept of reactor disposal in space; information on decontamination problems, potential missile damage, and reactor component dispersal from an accidental excursion; and to supply a large short burst of neutrons to external components The experiment rotated the control drums as fast as possible (about 100 times the normal rate) in the reactor.

This test provided the basic experimental information required for potential nuclear rocket accident analysis. The core suffered a thermally induced mechanical explosion--not a nuclear one! About 5-15% of the core vaporized in the test. One piece of the reactor pressure vessel weighing 44 kg was found some 457-533 m from the reactor.

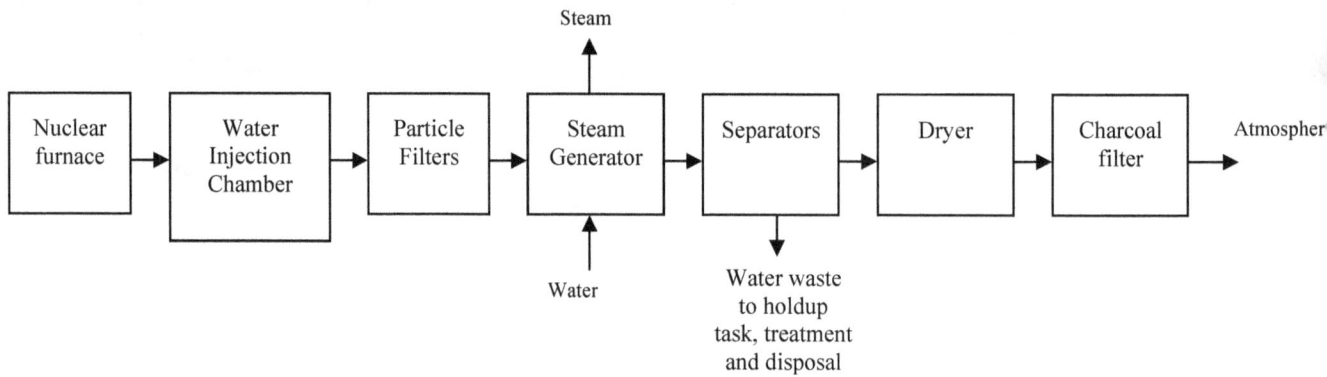

Fig. 8. Nuclear Furnace scrubber concept.

Special Precautions Needed At The Launch Pad

The requirement is to maintain the radiation dose levels below established health standards. Design options include not operating the reactor prior to launch (a zero power reactor test can be performed to verify the reactor is physical assembly correctly). This is to avoid radiation buildup in the reactor. Acceptance testing at the launch facility will be needed to ensure that all components are functional prior to mating with the launch vehicle. This could include cold flow testing; that is, testing where hydrogen is run through the engine for short periods of time to demonstrate that all valves and the turbopump are operational while maintaining the reactor shutdown.

The U.S. has launched one space reactor that was in a power generation configuration. This reactor, SNAP-10A, demonstrated the capability to launch a reactor without special radiation handling at the launch site. Further, nuclear fuels and reactors are transported around the country using well established containers and procedures. Sufficient design and operational experience exist to avoid transportation criticality accidents.

Launch Site Contamination Accidents--Fires

The primary requirement is to maintain the reactor subcritical without releases of hazardous radiation or radioactive materials during a fire. Design options relate to choice of materials and physical layout on the launch vehicle. For example, in case of an accident, it is more desirable to have the nuclear rocket in-line with chemical boosters rather than along side of them.

A series of propellant fire tests were performed as part of a project called Pyro to investigate the temperatures and duration of liquid propellant fires.[11] Theoretical data showed a peak temperature of 2,900 K for hydrogen-oxygen fires. The experimental data measured 2,500 K. This is below the melting temperatures of the nuclear rocket fuel, so that melting is not a problem. An analysis of the structural materials also indicate that melting is insufficient to cause a critical mass.

Solid propellant test show that they burn at approximately 3,000 K, with some chucks burning for up to 10 minutes. Again, the fuel melting temperatures are above the fire temperature. Using evaluations of the Lincoln Laboratory Experimental Satellites LES 8/9 that used a Titan III launch vehicle, the probability of an accident is 2 - 3 in a hundred. In a given accident, the probability of propellant chunks being in close proximity to the reactor is between one in a thousand and one in a million.

The conclusion is that the reactor can be designed not to melt or go critical in a launch pad fire. Detailed evaluations will be needed of particular nuclear thermal rocket and launch vehicle configurations.

Launch Site Contamination Accidents--Explosions

Here, the requirements are to prevent core compaction criticality and dispersal of radioactive materials. Design options are based on analysis from SP-100 nuclear power system where it was shown that the reactor would not go critical from the blast affects of launch vehicles. Similar design features can be built into nuclear rocket engines. Fragments may shear through the engine, but no fission fragment inventory exists within the core at this time. Therefore, no significant radiological risk from an explosion is projected. A major safety analytical and experimental program has shown that radioisotope generators are safe to launch;[12] NTPs, with their geometry and non-radioactive materials at launch, should be even less risk at launch.

Ground Nuclear Waste

The requirement is to minimize the amount of radioactive waste generated during the NTP program, especially long life waste. Detailed issues are addressed as part of a programmatic Environmental Impact Statement. NTP characteristics tend to minimize nuclear waste because of the very short operating times, measured in hours. Reprocessing of the fuel and burning the actinides can minimize/eliminate nuclear waste. This was demonstrated when NERVA fuel was reprocessed and reused.

Launch Approval

It is required that a formal flight safety review be completed with the approval of the Office of the President before nuclear power systems can be launched in the United States. This process, shown in Fig. 9, requires an independent review by the Interagency Nuclear Safety Review Panel that performs safety and risk evaluations.[13] The Panel provides the necessary independent risk evaluation that will be used by decision makers who must weigh the benefits of the mission against the potential risks. The agency, wanting to fly a nuclear powered payload than requests permission for flight to the Office of Science and Technology (OSTP). OSTP reviews the request and makes the launch decision; however, the Executive Office of the President makes the final decision if OSTP feels that is appropriate.

Crew Safety From Reactor Radiation

NASA crew dose guidelines for astronauts is 50 rem / year. Mars trips involve crew exposures to galactic cosmic radiation, Earth's radiation belts, solar radiation, and reactor radiation. Galactic cosmic radiation is continuous between 24 and 60 rem/year. Solar flares are stochastic short duration events with potentially high doses (>120 rem); the crew can be protected by a storm shelter for the limited duration of the events. Earth's belt radiation is minimized by limiting the amount of time spent there. The radiation to the crew from a NTP reactor is reduced by spacecraft geometry, local reactor shielding, hydrogen tanks and spacecraft shielding to levels of about one rem. For a typical NTP Mars trip, see Table 6, the radiation exposure levels for the crew is about 45 rem, of which the reactor contributes less than 3%.

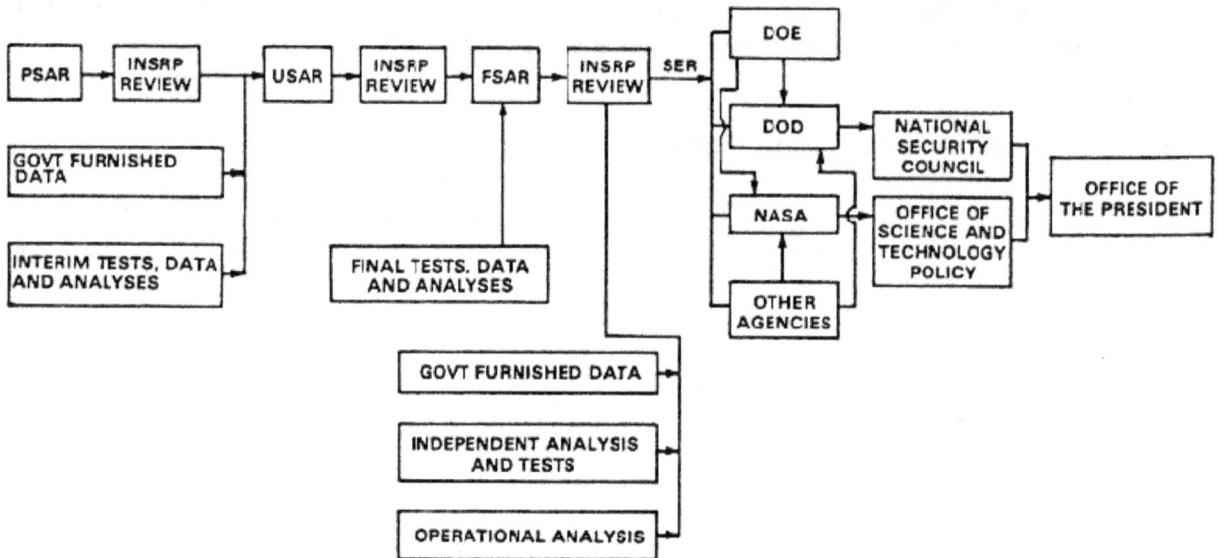

NOTE: FOR EXISTING NUCLEAR SYSTEMS, THE PSAR MAY NOT BE NEEDED.
PSAR = PRELIMINARY SAFETY ANALYSIS REPORT
USAR = UPDATED SAFETY ANALYSIS REPORT
FSAR = FINAL SAFETY ANALYSIS REPORT
SER   = SAFETY EVALUATION REPORT
INSRP = INTERAGENCY NUCLEAR SAFETY REVIEW PANEL

Fig. 9. Flight Safety Review and Launch Approval Process for space nuclear power systems.

Table 6. Typical Mars Mission crew radiation exposure.

Typical Mars Mission (rem	
Galactic	34
Solar flares (with storm shelter)	7.7
Earth radiation belts	1.5
Nuclear rocket	<1.1
Mars (30 days)	<1
Total	**45.3**

Turning now to launch and space operations, the questions in Table 4 will be addressed.

Criticality Prevention During Launch/Ascent Accidents

Requirements are for the system to remain subcritical for all credible launch/ascent accidents and to have no power operations until the system achieves its intended orbit or flight path trajectory. Design options include a built-in redundant shutdown subsystems with sufficient design margins in each system to ensure shutdown in case of a failure within either subsystem. NERVA was designed, in addition to its control drums, with neutron absorption wires in the core through the nozzle to further protect against launch criticality. Configurations can include in-core safety/shutdown rods or wires with locking devices and weak links. Command destruction of the reactor can be provided to ensure that debris from an accident terminates in an ocean.

Plans To Cleanup Contaminated Land Areas

If radioactive debris is deposited on land areas, it will be necessary to remove the material to designated storage sites. The approach here is mainly a preventative one. If an abort occurs near the beginning of the mission, the vehicle will likely land in the Atlantic Ocean. Based on Titan and Shuttle data, one failure in 57 flights of the solid rockets have occurred; however, no land impacts have occurred on other continents. The footprint from aborts later in the flight profile can be controlled by command destruct mechanisms to cause debris to fall into an ocean. Also, the reactor contains no radioactive fission products at launch. In the unlikely event of land debris impact, standard clean up organizations and mechanisms are in place. This includes the NEST Team (Nuclear Emergency Search Team).

Operation Below "Nuclear Safe Orbit" or "Sufficiently High Orbit"

"Nuclear Safe Orbit" (NSO) or "Sufficiently High Orbit" (SHO) refers to the acceptable reactor space storage location after use. The latter term, SHO, is now preferred. It means an orbital lifetime long enough to allow for sufficient decay of the fission products to approximately the activity of the actinides before reentry occurs. One design option is to initiate operations above the SHO for a given mission. However, for Mars missions and many others, it may be highly desirable to start below SHO. For these missions, provision can be made for on-board or external boost systems. Nuclear thermal propulsion stages can be throttled to ensure that the thrust vector is in an increasingly safe direction before accelerating to full propulsion power. The stage can be slowly rotated to average the thrust direction to safeguard against thrust nozzle misalignment failures. If fission gas retention is a problem at the higher temperatures, and correspondent higher rocket performance, the temperature can be reduced until the altitude is such that the fission gases are no longer a problem. If multiple nuclear rocket engines are included in a manned Mars mission, they would provide added safety in starting the nuclear propulsion stage below a SHO.

On-board devices have generally been used to boost low altitude satellites to higher orbits. This approach has been demonstrated on the U.S.S.R. RORSAT satellites. However, these sometimes fail. An external capability was being evaluated under a project called SIREN (Search, Intercept, Recover, Expulsion Nuclear) for boosting radioactive materials to higher orbit.[14]

A SHO is a function of the operating history. Typically, an orbital lifetime of 300 years has been used as the time for the radioactive materials from nuclear power plants to decay to safe radioactive levels. The orbital lifetime is a function of the ballistic characteristics of the spacecraft. Fig. 10 shows the orbital decay time as a function of altitude in terms of mass, drag coefficient, and cross sectional area. To achieve an orbital lifetime of 300 years corresponds to an orbital altitude above 400-500 nmi.[15]

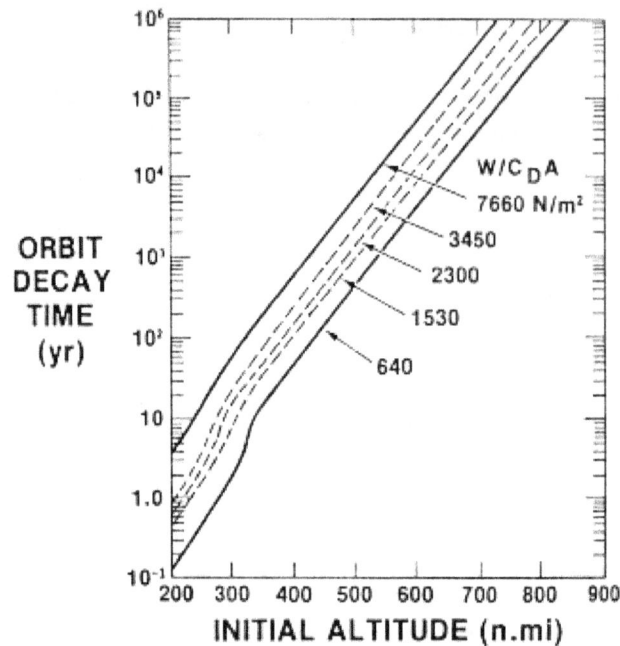

Fig. 10. Orbital decay times.

Near-Earth Space Contamination
The requirements include no significant additions of radioactive materials to the near Earth environment and protection of crews from exposures that exceed safety limits. During flight operations, insignificant amounts of fission products are expected to be released. These should mostly be in the form of the fission gases. As part of the flight environmental impact statement, an assessment will need to establish acceptable fission gas release levels. If a sensitive environmental area is being traversed, power and temperature can be reduced to maintain releases to background levels.

Affects On Other Satellites and Experiments
It will be necessary to avoid/minimize affects on other satellites. This can be accomplished as part of particular mission planning. Operations should generally be well away from other satellites; the radiation level exposures at other Earth satellites is a function of distance, power level and duration. Just from the desire to avoid collisions these should be negligible; however, power can temporarily be reduced if necessary in the vicinity of other satellites. During the limited operating time while leaving the vicinity of Earth (about 90 minutes), radiation sensitive sensors on other satellites will probably record the nuclear radiation from the reactor.

Final Disposal of Nuclear Engines/Prevention Of Nuclear Rocket Earth Impact
Final disposal of nuclear engines must be such that there is negligible probability of intersecting or passing within the close proximity of Earth. From Mars, since NTR reuse is not planned by the Synthesis Group, it would have been placed in a deep space orbit that would not intersect the Earth. The NTR stage can be ejected after the Mars burn or mid Course correction used to return the manned spacecraft to Earth; the NTR is not planned to be used in spacecraft Earth capture or achieving Earth orbit. From the Moon, if reuse was not planned, the NTR can easily be placed in a deep space disposal orbit. For a nuclear tug, it will eventually be disposed of either above a Sufficiently High Orbit or in deep space, not back to Earth.

Safety Questions Summary
Nuclear thermal propulsion can be designed to operate safely if safety standards are defined at the initiation of any nuclear thermal propulsion program and continuously monitored for compliance. Design and operational solutions to meet these standards have been addressed in previous programs, such as NERVA. The solutions depend on particular concepts and their intended missions. However, after reviewing a wide range of

questions related to safety, there were no questions that did not appear to have practical design/operational solutions.

Reliability

Emphasis was placed on reliability in the NERVA flight engine development program. In fact, because the NERVA engine was being designed for manned trips to Mars, it was considered the key design requirement. As stated by Milton Klein, the Program Director,

> "The major design criteria for the NERVA engine development program shall be reliability and the achievement of the highest probability of mission success. Next in order of importance must be performance as measured in terms of specific impulse. Then the engine design should attempt to keep the overall weight as low as possible."[16]

This led to a reliability requirement of 0.997 at 90% confidence level This very high reliability requirement resulted in significant design efforts and the development of a methodology to show that the engine design could achieve this high reliability and confidence level. Future thermal nuclear rockets that are being designed for manned missions are expected to have similar very high reliability requirements. The NERVA methodology provides a good base to demonstrate that the reliability goals are attainable.

Cost, Schedule and Man-Rating

When specifying requirements for future nuclear thermal propulsion systems, it is important to consider the cost and schedule implications. Specifying performance above the minimum to do the missions can result in large additional development and verification costs! On the other hand, the cost and schedule should not compromise the minimum mission, safety and reliability requirements that the developed rocket needs to meet.

Lower thrust engines, such as a 66 kN (15,000 lb$_f$), greatly reduces the size of the testing facilities needed and the cost of manufacturing test articles. Also, very significantly, existing launch vehicles can be used to flight qualify a 66 kN (15,000 lb$_f$) rocket.

Designs should be favored that the design margins can be established at the component level. Validation costs increase approximately by factors of ten as one goes from elements to components to subsystems to systems.

Also, the earlier in the schedule that key issues are resolved, the less cost and schedule risk. Therefore, it will be less costly and less risky to select design concepts that can be validated at the component level rather than system level.

Competition is one of the best ways to control costs. Compete development efforts at the component level. At the systems level this is usually too expensive. If the project is organized to bring in new components as they are demonstrated, competition can be maintained throughout the development program at reasonable costs. This also allows the incorporation of new ideas into the development process.

Move into systems level validation only when high confident components are available. Systems testing is relatively expensive; if the designs are such that there is little uncertainty at this level, the systems testing becomes mainly a validation that the component specifications were correct and everything works as designed.

Man-rating the nuclear rocket for missions such as Mars involves space experience. Getting this experience in useful missions is important. These missions should be ones that can be performed at a reasonable cost and furnish new scientific data. The outer planet exploration program offers many opportunities for both the nuclear rocket experience and to perform more demanding science missions. Many of these missions can use Delta II, Atlas IIAS, Titan III or Titan IV existing launch vehicles.

Possible Future Nuclear Thermal Propulsion Requirements

Based on the criteria presented above, a possible set of requirements for a future nuclear thermal propulsion development program are:[17]

- Thrust level of 66 kN (15,000 lb$_f$) for each engine. With a cluster of four engines, the Mars crew mission can be accomplished. Multiple engines enhance crew safety and mission reliability and meets a fail-operational, fail-safe criterion for the propulsion module. It provides a means to successfully complete the mission even with the loss of one or possibly more engines. With multiple engines, each individual engine does not have to be as reliable as a single engine trying to do the whole mission In addition, the thrust level is compatible with the outer planetary missions. For example, it reduces the critical nuclear electric propulsion times for a Pluto orbiter mission by one-third. Development costs and schedule are reduced by requiring smaller facilities, reducing the cost per engine by about one-third, and building on existing fuels, facilities, and infrastructure.
- Specific Impulse greater than 8.1 km/s (825 seconds). This is sufficient for both the piloted Mars and outer planetary orbiter missions. It reduces development cost and risk by building on the carbide bead fuel already demonstrated in the Rover/NERVA programs at the temperatures and lifetimes needed. Many engine concepts can use this fuel. If the Russian fuel is available and proves to do what is claimed, the recommended specific impulse can be raised to 8.8 km / s (900 s) with little additional risk. Or, if an improved fuel is available in time, this can also be used.
- Thrust/weight greater than 3 for the propulsion module. This provides acceptable mission performance. In 1972, LANL designed an engine at 72 kN (16,000 lb$_f$) thrust with a thrust/weight ratio of 2.8. Improved concepts and components means that thrust/weight ratios of greater than 3 are achievable at reasonable risk. More than one perigee burn can be used to provide mission flexibility if the thrust/weight goals are not achieved.
- Burn time of 1.25 hours. This is sufficient to do the most stressing Mars mission assuming that the launch vehicle is in the 200-250 tonnes class, and that nuclear thermal propulsion will be used to escape Earth, perform Mars orbit insertion, and return the crew to Earth. The NERVA NRX-A6 testing demonstrated that the fuel exists to meet this requirement. At least 10 starts should be specified to cover perigee propulsion missions. This will cover all phases of the Mars missions with some margins. NERVA XE testing, with 28 restarts, demonstrated that this is achievable.
- Regenerative cooled nozzle should be less than 10:1. This reduces the facility costs and is achievable with today's technology.
- For manned Mars missions use a four engine module for crew safety. This meets the criterion of fail-operational, fail-safe; the propulsion module degrades gracefully; removes propulsion risk as a major concern in completing the mission; and provides flexibility in operations (examples are: (1) if an engine fails on leaving Earth orbit, an additional perigee burn can be performed, and (2) only one operational engine is needed to return the crew to Earth from Mars orbit).
- Reliability of propulsion module greater than 0.999. This requires an engine mission reliability around 0.95, which is about what has been demonstrated with chemical rockets.
- Afterheat removal minimizing pulse cooling. Each engine should have a built-in capability to remove the afterheat to prevent core meltdown in case of a malfunction. As an example, this can be accomplished using a combination of an emergency cooldown tank and radiator. The mass saving should be more than that currently allocated for pulse cooling afterheat removal.
- Design to avoid non-independent failures. Non-independent failures should be eliminated (if possible). The major one identified to date is a turbopump failure. Proper design and orientation can avoid this. The use of the normal reactor control drum speeds prevents the core from catastrophically disassembling.

Chapter 3

ROVER/NERVA Programs

Overview

As early as 1946, in the infancy of the development of nuclear reactors, the potential of fission energy for rocket propulsion was proposed.[1] The fuel's high energy density and performance efficiency spurred this interest in direct nuclear thermal propulsion. In 1955, a program was initiated to pursue the development of a solid-core, hydrogen-cooled reactor in which the exiting gas expanded through a rocket nozzle and is discharged to space. This effort was technically a success; however, changing national priorities away from going to Mars following the successful Apollo program led to the programs termination in 1973.

Between 1955 and 1973, the United States pursued nuclear rocket research and development efforts under the Rover/NERVA (Nuclear Engine for Rocket Vehicle Application) programs. These programs had two primary elements:

> The Rover program at the Los Alamos National Laboratory performed advanced technology development including establishing the basic reactor design, leading the fuels development efforts, and establishing reactor testing capabilities.

> The NERVA program used the industrial team of Aerojet and Westinghouse to further the technology towards a flight system. The program included demonstration of a prototype flight reactor and rocket engine system

Reliability was a key requirement of the NERVA engine development program. "The major design criteria for the NERVA engine development program shall be reliability and the achievement of the highest probability of mission success. Next in order of importance must be performance as measured in terms of specific impulse. Then the engine design should attempt to keep the overall weight as low as possible."[2]

This led to a reliability requirement of 0.997 at 90% confidence level Other requirements were:
- A minimum specific impulse of 7,450 m/s (760 seconds)--about twice that of the best chemical rocket engines,
- The ability to operate over a wide range of power conditions,
- Engine start-up without external energy sources, a so-called bootstrap start,
- Programmed closed-loop control,
- Engine duration at full thrust of 10 hours, with restart capability, and
- Thrust of 337 kN (75,000 lb$_f$). [The initial requirement was for 245 kN (55,000 lb$_f$).]

Fig.1 shows the various nuclear reactor test series that were part of the Rover/NERVA Programs; Fig. 2 summarizes the test articles in the various test series.[3] The purpose of each of the test series will be discussed in the following paragraphs.

Research reactors - named Kiwi - were used to establish basic nuclear rocket reactor technology and to develop sound design concepts. The Kiwi series of reactors demonstrated high temperature fuels and were the first reactors to operate using stored liquid hydrogen (LH$_2$). Reactor control was automated using drums in the reflector.

Thermal-hydraulic vibrational problems occurred early in the Kiwi program. These problems were resolved following several cold flow and experimental reactor tests. The Kiwi program culminated in the Kiwi-B4E reactor operating at over 1,890 K for 11.3 minutes and at 2,005 K and 937 megawatts (MW$_t$) for 95 seconds.

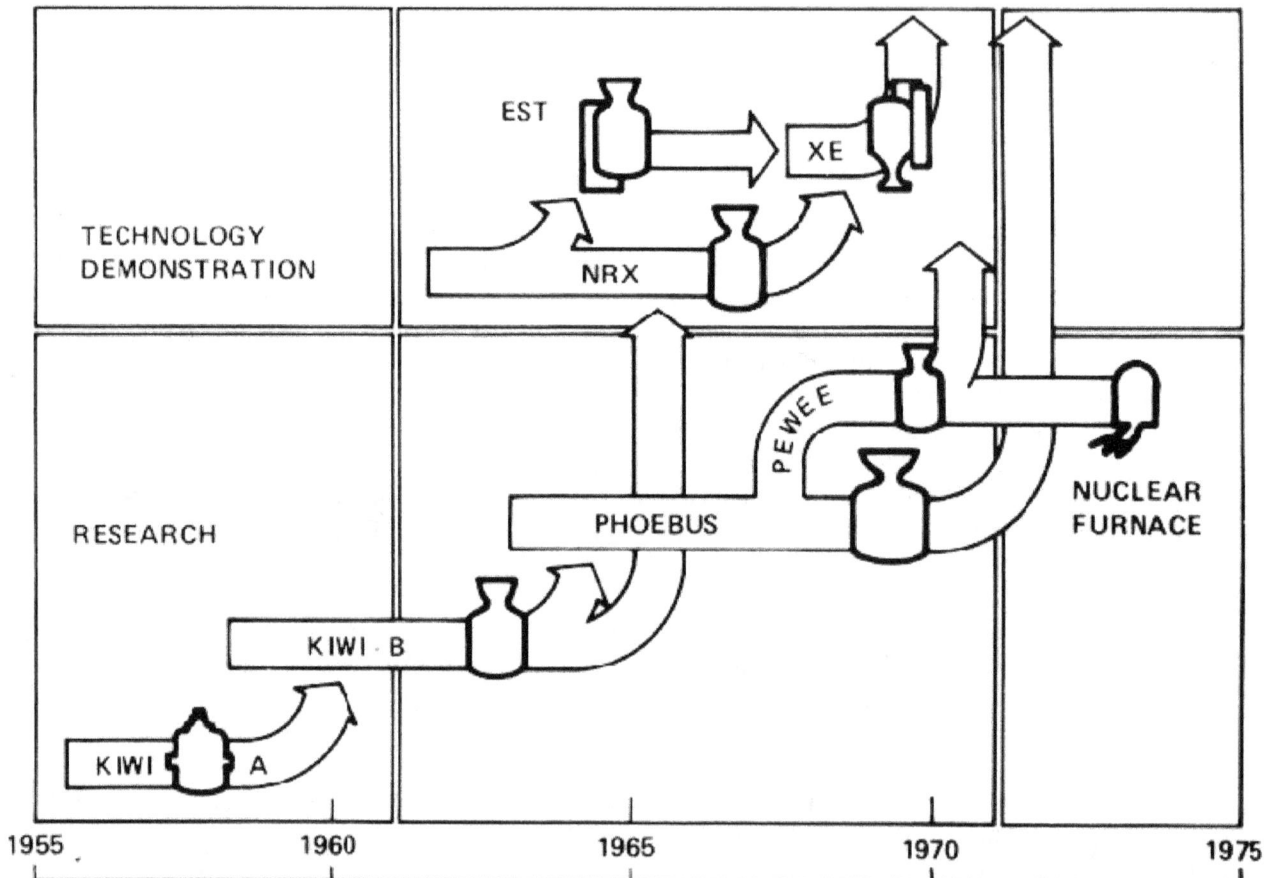

Fig. 1. Nuclear rocket reactor development series. *From G. H. Farbman and R. E. Thompson, AIAA Paper No. 75-1261.*

The Kiwi reactors success led to the NERVA program that developed the NRX series of developmental reactors. The NERVA program goal was to demonstrate a specific impulse of 7,450 m/s (760 seconds) for 60 minutes in a 1,100 MW reactor. The thrust level was 245 kilonewtons (kN) [55,000 Ib$_{force}$]. The reactor objectives were exceeded in the NRX-A6 test, which ran for 62 minutes at 2,220 K and 1,100 MW$_t$ with only 11¢ reactivity loss. The engine objectives were met in the XE test.

Under Rover, an additional series of research reactors, called Phoebus, were developed. The objectives were to demonstrate increased performance including: increasing the specific impulse to 8,085 m/s (825 seconds), increase power density 50 percent, and increase the power level to 4,000-5,000 MW$_t$. Success was demonstrated in the Phoebus-IB and Phoebus-2A tests. The Phoebus-2A test operated for over 12 minutes above 4,000 megawatts-thermal and reached a peak power of 4,100 MW$_t$.

The Pewee and Nuclear Furnace (NF) reactor series objectives were to demonstrate still higher temperature and longer life fuel elements. Pewee-I ran at 2,555 K and 514 MW for 40 minutes. NF-I ran at 2,450 K and 54 MW for 109 minutes.

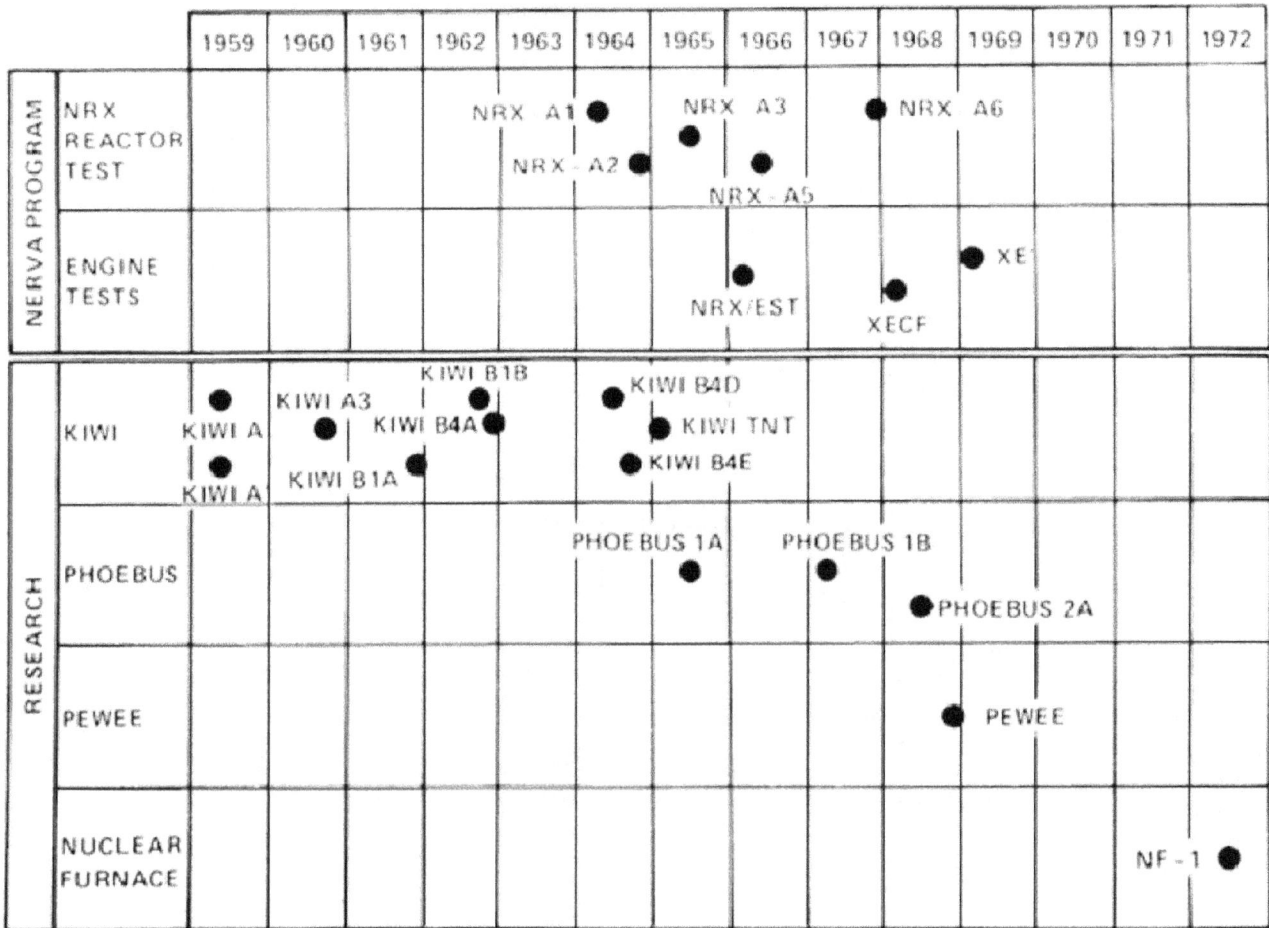

Fig. 2. Major tests accomplished during the Rover Program. *Courtesy of Los Alamos National Laboratory and Westinghouse Electric Co.*

The NERVA engine development test program was the first and only program that ran a nuclear rocket in a downward firing rocket configuration. An experimental engine was designed and tested that: (1) determined system characteristics during startup, full power, and shutdown conditions; (2) evaluated control concepts; and (3) qualified engine test stand operations in a downward firing configuration with simulated altitude conditions resembling operation in outer space. Additional reactor and engine development objectives were met or exceeded in the NRX/EST and XE test programs. XE was a prototype flight engine system. During testing in the simulated space environment, it performed some 28 starts and restarts. This engine included non-nuclear flight components that were demonstrated along with a flight-type reactor.

The NERVA program had initiated flight engine system designs; however, the program was terminated prior to this being completed. The program objectives were expanded to develop a suitable flight design that could operate for 10 hours and 60 cycles with a reliability of 0.997. The full-flow topping cycle where the propellant used to drive the turbopump is returned to the engine before the reactor inlet had been selected for the engine design.

Throughout the reactor development efforts, emphasis was to increase the reactor outlet temperature (since the specific impulse is proportional to the square root of the temperature) and the operating time of the reactor. The success of this part of the Rover Program is depicted in Fig. 3.[4] (XE Prime, though operated late in the development program, actually used older reactor components since the objectives were to demonstrate engine behavior rather than reactor performance.) Temperature levels at the core or fuel exit (T_{FE}) of over 2,500 K and operating times over two hours were demonstrated. The cumulative time-at-power is shown in Fig. 4. As the designs were perfected, there was an accelerated increase in the time-at-power.

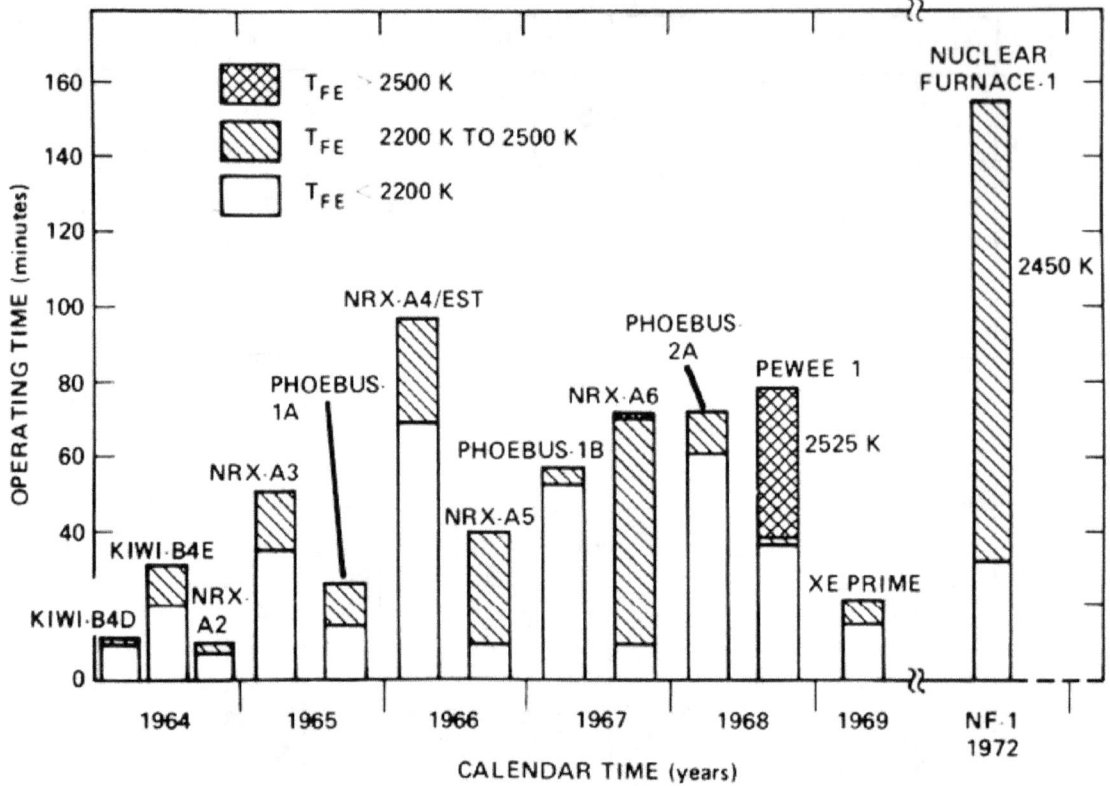

Fig. 3. Temperature performance demonstrated during rocket reactor tests. *After, David S. Gabriel, 1972.*

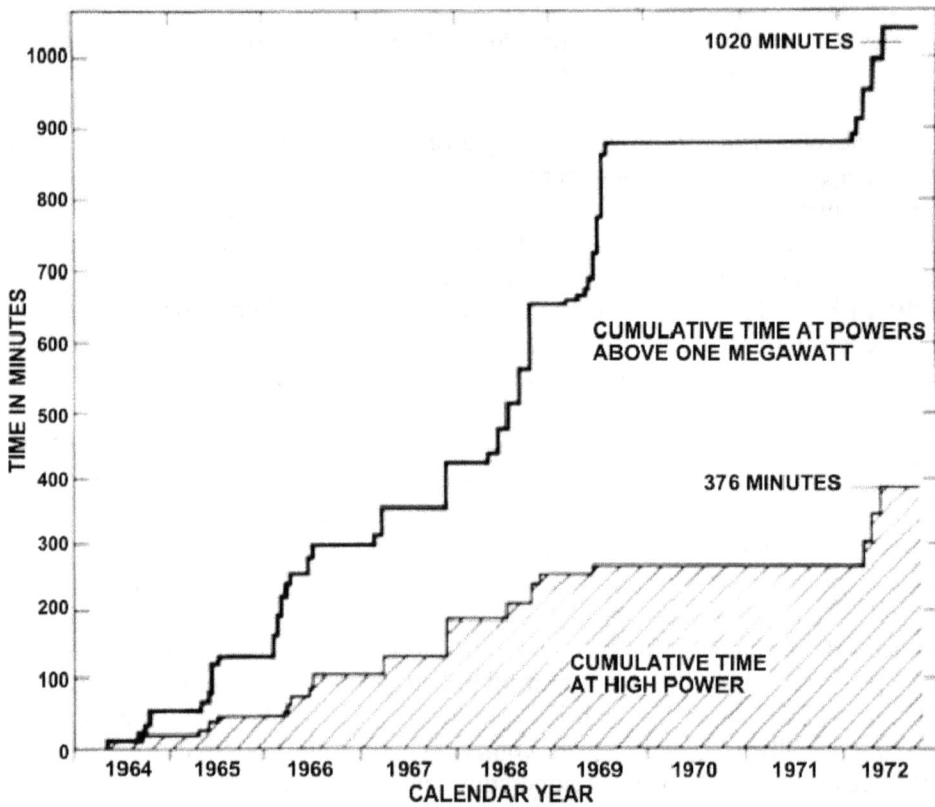

Fig. 4. Cumulative time-at-power during rocket reactor testing. *After David S. Gabriel, 1972.*

BOOK 2
NUCLEAR THERMAL PROPULSION SYSTEMS

Major Rover Program events are summarized in Table 1. The decision to proceed with the program was made in 1955, following several years of studies, and the first reactor was tested in 1959. Considering that a whole infrastructure had to be designed and constructed, progress was quite rapid. This included constructing test cells and hot cells for assembling than disassembling highly radioactive reactors. Industrial contractors were brought into the program in 1961 and a reactor in-flight test program was started. The in-flight test program lasted two years, at which point it was decided to concentrate on engine development. By the end of 1964, the Kiwi program had demonstrated full-power operation, overcoming vibrational problems. The first fueled demonstration reactor, NRX-A2, was run in 1964 at 1,100 MW$_t$ for about five minutes. Then, in 1965, as a safety test, a reactor was intentionally (but with great difficulty) put on a very fast transient to observe what would happen when it destroyed itself. Mechanical disintegration, not a nuclear explosion, occurred.

In 1965, a new class of reactor development with a power level in the 4,000-5,000 MW$_t$ range was initiated. In 1966, the first prototype breadboard engine system, called the NRX/ EST, was tested. Many successful startups and shutdowns were achieved. By 1967, the final reactor in the NRX series, NRX-A6, had operated, meeting the established development engine reactor goals. Progress continued with, in 1968, the highest power steady-state reactor ever built being operated in the Phoebus 2A test. It performed above 4,000 MW$_t$ for 12 minutes. In addition, in 1968, the Pewee-1 reactor test established a core power density record. The year 1969 witnessed the downward firing of the first prototype nuclear rocket engine, including operating in a simulated space environment. Prior to this time, all the rocket test reactors had operated in an upward firing position. The configuration tested was similar to a flight engine.

However, in 1969, a political decision was made to suspend production of the Saturn V launch vehicle. The consequences were significant for the nuclear rocket program. The Saturn V vehicle was to be the prime launch vehicle for the nuclear rocket. This cancellation heralded a decision to abandon the human exploration of Mars as a direct follow-on to the Apollo lunar landings.

In 1972 a final space reactor test was conducted in a nuclear furnace (NF) system. This test demonstrated up to 109 minutes operation at power densities of approximately 4,500 MW$_t$/ m^3 and temperatures of over 2,500 K. Then, in 1973, the nuclear rocket program was terminated--a technically successful program! However, it fell victim to changing national space program priorities.

Table 1. History of the Rover Program

1955	Following several years of nuclear rocket studies, nuclear rocket program initiated as Project Rover at Los Alamos National Laboratory. Concept to be pursued was solid core, H_2 cooled, reactor expanding gas through a rocket nozzle.
July 1959	First reactor test, Kiwi-A, tested at 70 MW for 5 minutes.
Oct 1960	Proof-of-principle tests (Kiwi-A series of 3 reactors) completed.
July 1961	Industrial contractors (Aerojet-General for rocket engine and Westinghouse Electric Corporation for reactor) selected to perform rocket development phase. Reactor in-flight tests (RIFT) program initiated.
1963	RIFT program canceled.
1961–1964	Kiwi-B series of 1,000 MW reactor tests included five reactors plus several cold-flow unfueled reactors to resolve vibration problems and demonstrate design power.
May–Sept 1964	First full power test, Kiwi-B4D, at design power with no indications of core vibrations. Also demonstrated restart capability.
Sept 1964	NRX-A2, first tests of the NERVA (NERVA = nuclear engine for rocket vehicle application) reactor, reached full power of 1,100 MW for about 5 minutes.
Jan 1965	Kiwi-B-type reactor deliberately placed on fast transient to destroy itself as part of safety program.
June 1965	The prototype of a new class of reactors, Phoebus-1A, was run at full power for 10.5 minutes.
Mar 1966	The NRX/EST, first rocket engine breadboard powerplant, operated at full power (1,100 MW) for 13.5 minutes.
Dec 1967	The fifth fueled NRX reactor in the NERVA engine series exceeded the design goal of 60 minutes at 1,100 MW.
June 1968	The Phoebus-2A—the most powerful nuclear rocket reactor ever built—ran for 12 minutes above 4,000 MW.
Dec 1968	Pewee set records in power density and temperature operating at 503 MW for 40 minutes at 2,550 K and core power density 2,340 MW/m^3.
Mar 1969	The first down-firing prototype nuclear rocket engine, XE-prime, was successfully operated at 1,100 MW.
1969	Saturn V production suspended—prime launch vehicle for NERVA.
June 1972	In the 44 MW Nuclear Furnace (NF-1), fuel was demonstrated at peak power densities of approximately 4,500 MW/m^3 and temperatures up to 2,500 K for 109 minutes.
Jan 1973	Nuclear rocket program terminated. Judged a technical success but changing national priorities resulted in cancellation decision.

Reactor Development

Reactor development goals included maximizes the core exit temperature of the hydrogen propellant, increasing operational lifetimes (from the initial one hour to ten hours operation), minimizing hydrogen corrosion of the nuclear fuel, and minimizing fuel breakage from vibrational and thermal stress. Nuclear rocket reactor schematic and cutaway views are presented in Fig. 5. The reactor was an epithermal, graphite moderated, hydrogen-cooled reactor, using enriched (93.15%) uranium-235 as the nuclear fuel.[5]

Fuel elements were hexagonally shaped, with 19 cooling channels per element (see Fig. 6).[6] The reactor core was supported by a cold end support plate with a structural tube arrangement tied to the support plate. Power flattening was achieved by varying the fuel loading. Hydrogen flow distribution throughout the core was controlled by orifices in the core inlet plate. As seen in the schematic, the reactor core was surrounded by a beryllium reflector which contained rotating drums for power control. These drums controlled the reactor reactivity by positioning a neutron absorber material (boron carbide).

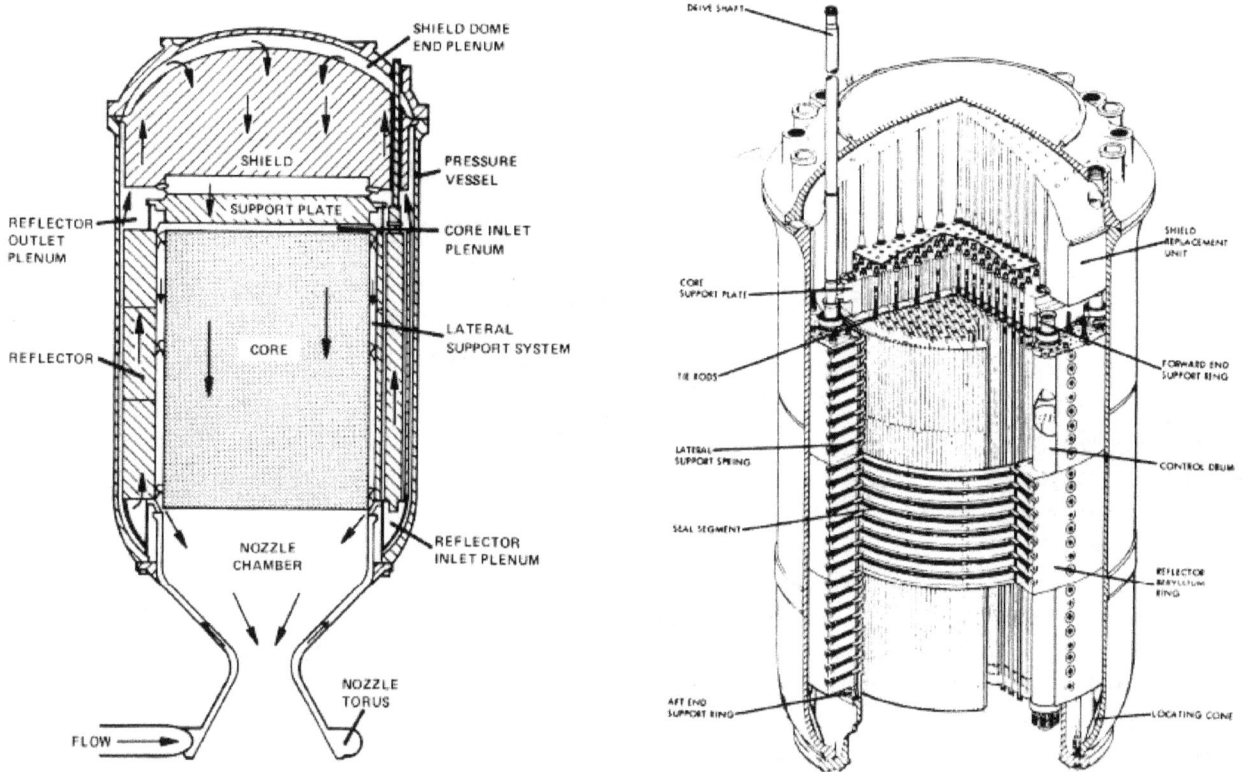

Fig. 5. Schematic (left) and cutaway (right) of Rover nuclear rocket reactor. *From Westinghouse Astronuclear Laboratory, November 1967.*

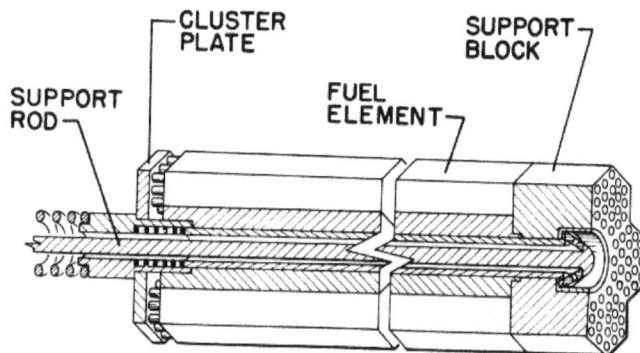

Fig. 6. Fuel element cluster employed in most of the later Rover reactor designs. It consists of six, full-length, hexagonal fuel elements supported by a centrally located tie rod. Each extruded graphite fuel element had 19 cooling channels. *From Daniel R. Koenig, 1986.*

The performance of some of the more significant Rover reactor tests is summarized in Table 2. Tests ranged from about 500 to 4,100 MW_t and involved hydrogen flow rates of about 19 to 119 kg / s. The difference between the chamber temperature and the fuel exit average temperature depended on the amount of propellant flow used to cool the periphery. In many of the experimental reactor tests, a large amount of periphery propellant flow was used; the fuel exit temperature rather than the chamber temperature is really more indicative of projected flight performance. Flight reactor designs were configured to minimize this excess of peripheral flow. Pewee-1 achieved the highest exit temperature with an exit temperature of 2,556 K.

Table 2. Performance of various Rover reactor systems.

	KIWI-4BE	NRX-A6	Phoebus-2A	Pewee-1
Reactor power (MW)	950	1,167	4,080	507
Flow rate (kg/s)	31.8	32.7	119.2	18.6
Fuel exit average temperature (K)	2,330	2,472	2,283	2,556
Chamber temperature (K)	1,980	2,342	2,256	1,837
Chamber pressure (MPa)	3.49	4.13	3.83	4.28
Core inlet temperature (K)	104	128	137	128
Core inlet pressure (MPa)	4.02	4.96	4.73	5.56
Reflector inlet temperature (K)	72	84	68	79
Reflector inlet pressure (MPa)	4.32	5.19	5.39	5.79
Periphery and structural flow (kg/s)	2.0	0.4	2.3	6.48

Nuclear fuel improvements were a major technology development effort. The original Kiwi-A and Kiwi-B through the Kiwi-B4D reactors used highly enriched uranium dioxide (UO_2) fuel, extruded in a carbide matrix. The particle size was approximately 4 micrometers in diameter and its density was approximately 10.9 kg / m^3. Melting temperature of this fuel is 2,683 K. The demonstrated operating limit of this design series was 20 seconds at 2,127 K. Major engineering problems were dimension control and fuel erosion. In the 1,875 to 2,275 K temperature regime, the UO_2 reacted with the carbon, converting to UC_2 with the evolution of CO and loss of carbon from the fuel element.

The Kiwi-B4E, Phoebus, Pewee, and NRX-A reactor tests replaced the UO_2 fuel with beaded UC_2 particles with pyrolitic graphite coatings as protection against oxidation and storage (illustrated on the left side of Fig. 7). Fuel particle size was in the range of 50 to 150 micrometers in diameter with a 25 micrometer thick coating. in addition, each fuel elements used a graphite matrix with niobium carbide (NbC) coatings to protect against hydrogen corrosion. Later reactor designs replaced the NbC with zirconium carbide (ZrC). This fuel demonstrated one hour operation at temperatures between 2,400 K and 2,600 K. The major problems encountered with the beaded UC_2 particle coated fuel involved the large difference between the coefficients of thermal expansion of the graphite matrix and the NbC coatings. Excessive carbon loss after one hour occurred in the 2,375 to 2,575 K temperature range.

A composite (U, Zr)C particles coated with ZrC (illustrated on the right side of Fig. 7) was tested in the nuclear furnace (NF-l). This fuel performed for 109 minutes at 2,450 K at peak power densities of approximately 4,500 MW_t / m^3. Projections indicated that with matched thermal conductivity this fuel would be good for four to six hours at 2,500 to 2,800 K. The major problem observed in the NF-l test was still some cracking from radiation damage. Also included in the NF-l experiment were several fuel elements of small, pure carbide particles (U,Zr)C. These fuel elements exhibited extensive cracking at the power levels occurring in the NF -1 experiment.

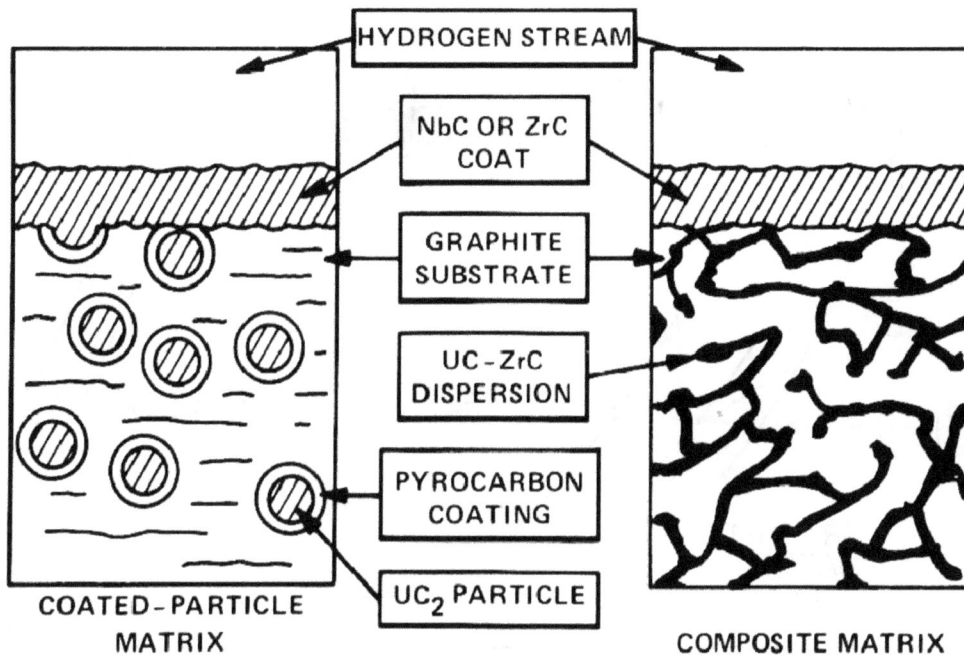

Fig. 7. Comparison of Rover Program fuel elements. *From David S. Gabriel, 1972.*

A problem encountered with all the fuel configurations was called "mid-range corrosion." This is illustrated in Fig. 8. The fuel operated at much higher temperatures toward the chamber end of the core than the front end. To accommodate this, the fuel was processed to accept the high chamber end temperatures. The front end of the core had low mass loss rates because of the low temperatures found in that region. The problem was in the mid-range. In the mid-range there tended to be the highest mass loss rates because temperatures were appreciable and the neutron flux level was high -- the temperatures were appreciatively different from the high end region for which the fuel was processed. On the hot end of the core, the temperatures matched the processing temperatures and the neutron flux levels tended to be low.

Also seen in Fig. 8 is the advantage of the zirconium carbide.[7] Even more advantageous is the use of a composite fuel coated with ZrC.

Even greater performance gains are projected by changing the fuel to a carbide. Fig. 9 illustrates the anticipated lifetimes at various operating temperatures for three fuels of interest.[8] If a ten hour life is desired, the reactor would be limited to operation around 2,200-2,300 K with a graphite matrix fuel; while with a composite fuel, this could go up to almost 2,400 K; and with a carbide fuel, up to approximately 3,000 K. For one hour operation, the graphite matrix fuel could be operated at over 2,500 K, the composite at 2,700 K, and the carbide possibly as high as 3,300 K. Thus, as seen in Fig. 10, nuclear thermal rocket engine specific impulse can possibly exceed 1,000 s performance.

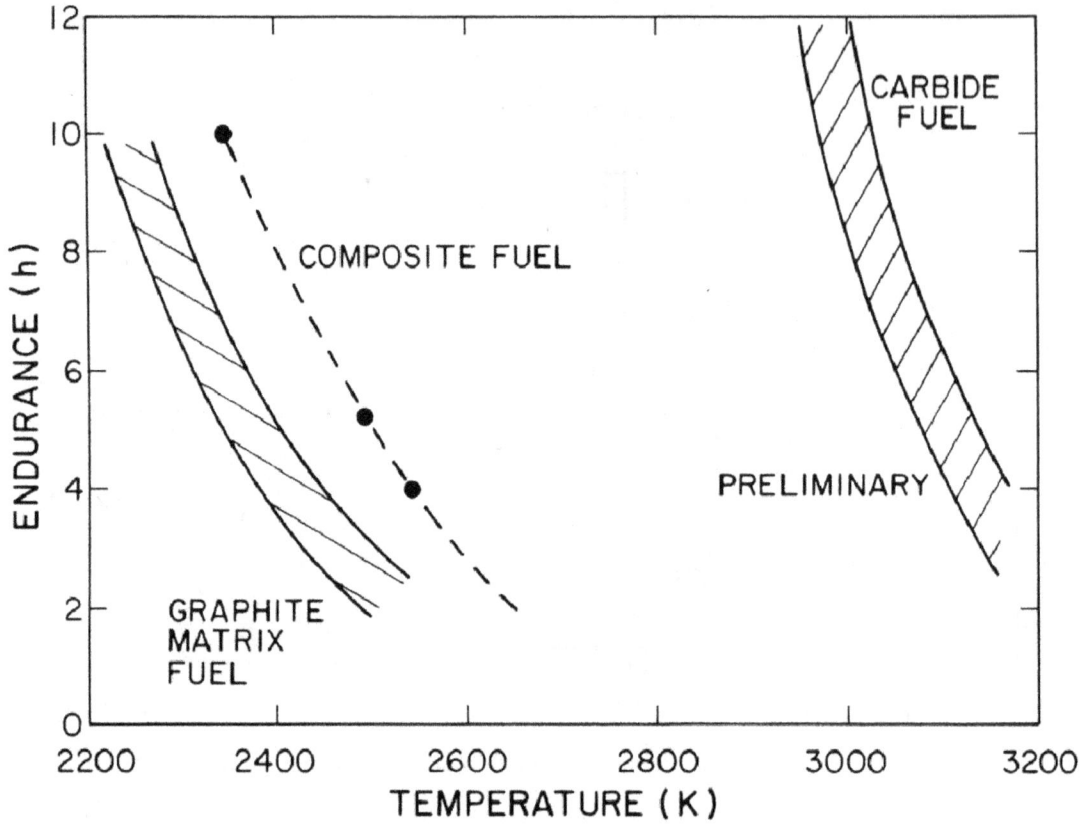

Fig. 8. Fuel element environment and mass loss from Pewee and NF -I fuels. *After Luther L. Lyon, September 1973.*

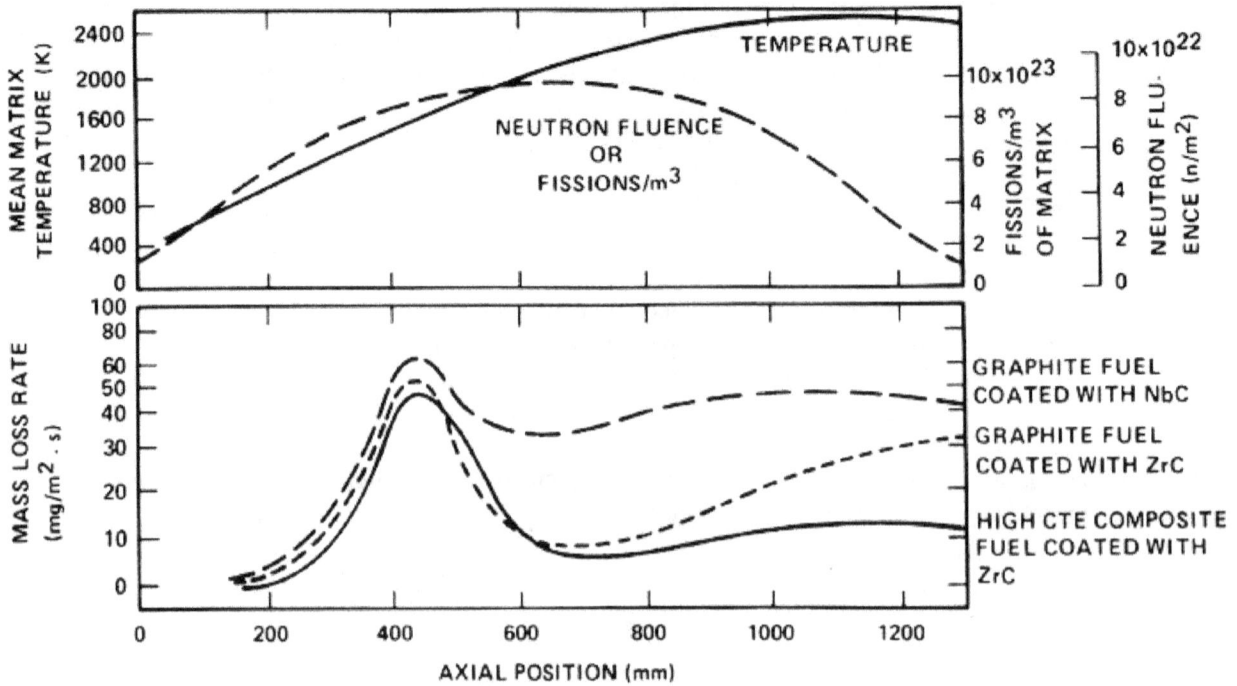

Fig. 9. Anticipated lifetimes of various Rover reactor fuels. *From David S. Gabriel,1973.*

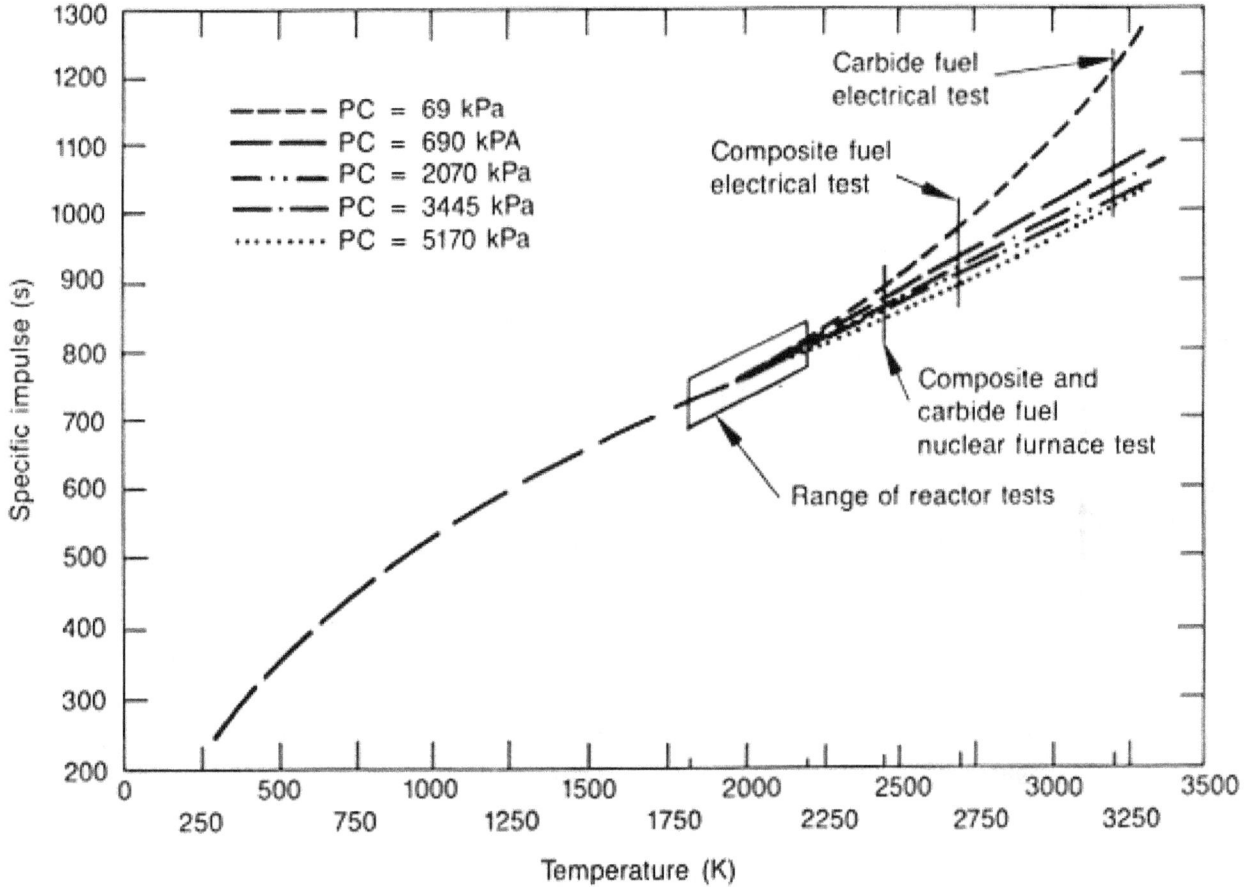

Fig. 10.   Specific impulse versus exhaust nozzle temperature for nozzle area ratio 300.   From *Ramsthaler and David a. Baker,*

Fig. 11 is another way of depicting the progressive improvements achieved in fuel element development.   During nuclear rocket reactor operation, hydrogen corrosion can cause lost of fuel from the core.   Control drums in the reflector are used to adjust the core reactivity to compensate for this fuel lose.   Starting with the NRX-A2 and NRX-A3 reactors normalized to unity, by the end of the NRX series the fuels had improved to the point that there was less than a third of the relative corrosion rates as had occurred at the beginning of the NRX series.   Pewee-1 demonstrated even significantly greater improvement in reducing fuel corrosion.   The development fuel for the Pewee-2 (a system which was never constructed) was projected to reduced fuel corrosion to less than one-tenth the relative corrosion rate.

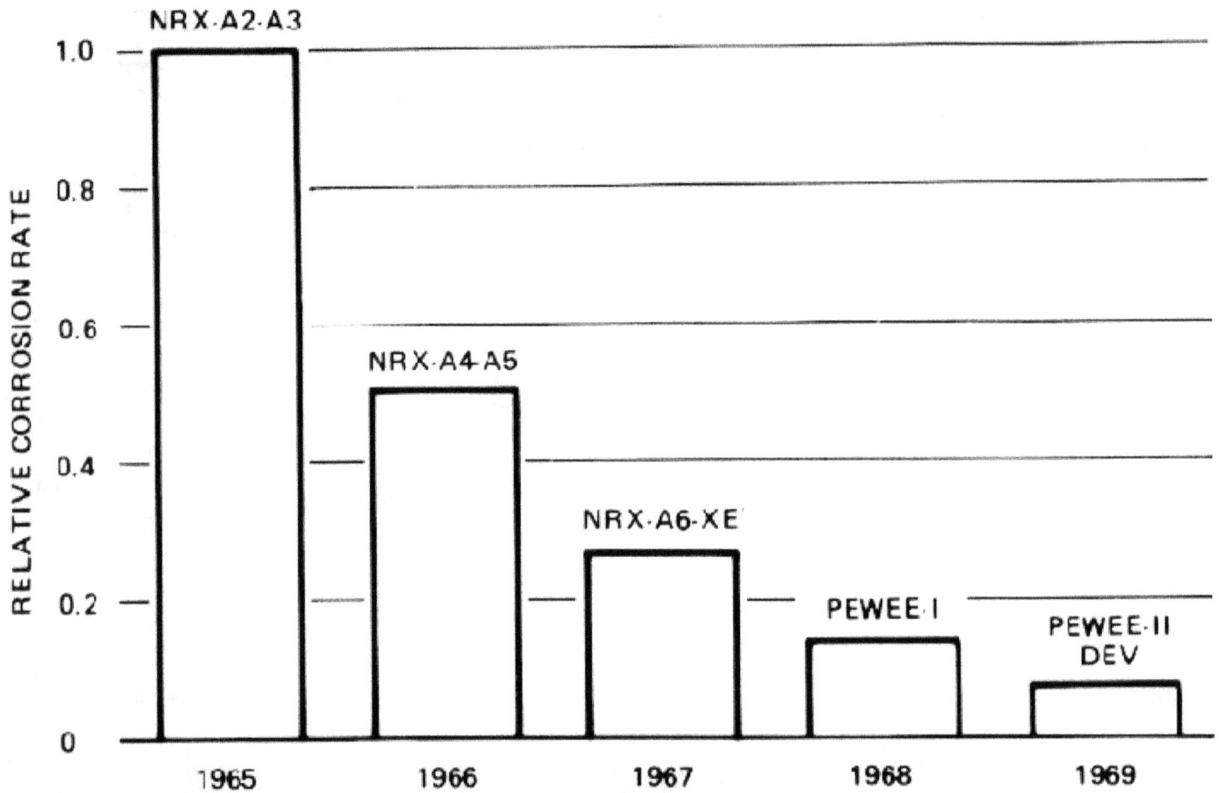

Fig. 11. Program in reducing hydrogen corrosion of fuel elements. *Courtesy of Los Alamos National Laboratory and Westinghouse Astronuclear Laboratory.*

Fuel structures were another significant problem in the Rover reactor development. The early Kiwi series quite literally fell apart from vibrations. These vibrations were induced by thermal-hydraulic interactions. By the end of the Kiwi series the redesigned structural support system demonstrated adequate performance. Further improvements in the structural system continued as summarized in Table.3. The structural anomalies that occurred were identified after the tests and these anomalies did not result in a termination of power. At the end of the NRX-A6 tests, some cracking was found in the beryllium reflector ring, in support blocks, in the peripheral deposit cups, and in one tungsten cup. These cracks were believed due to thermal gradients.

Table 3. Summary of NRX reactor structure anomalies.

Source	Significant Structural Anomalies	Phenomenon Causing/ Allowing Occurrence	Resulting Technique, Criteria or Geometry Change (Corrective Action)
NRX-A1	1. Control drum binding	1. Deflection of control drum and outer reflector due to thermal gradients	1. Complete deflection analysis of all interacting components for all reactor conditions indicated; increase bore diameter in outer reflector
	2. Pressure vessel-outer reflector support ring interference	2. Deflection due to thermal gradients	2. Deflection analysis for startup, steady state and shutdown required.
	3. Reflector cracks	3. Contraction due to thermal gradients on nozzle end ring	3. Finite element analysis technique was final technique prior to A6
NRX-A2	1. Aluminum barrel tab bent	1. Combination of barrel pre-test axial movement, shrinkage, and inner reflector axial contraction	1. None indicated, not considered serious
NRX-A3	1. Support block cracks, aft c' bore fillet to inner flow holes missed on A2 same cond.	1. Stresses due to thermal gradients and mechanical loads at termination of startup ramp	1. Initiated first finite element computer analysis study (This technique was developed extensively)
	2. Control drum rubbing	2. Deflection of sector and drum due to thermal gradients during cooldown after flow initiation	2. Transient analysis required, drum material removed
NRX-A4	1. Cracking of support blocks fillets also peripheral and axial cracks	1. Stresses due to thermal gradients	1. Reduction of start-ramp on A5, also planned test holds, also partial lobes removed, c'bore radius was increased
	2. Flaring and splitting of liner tube ends	2. Excessive growth due to overheating	2. Reduce liner tube length
	3. Light rubbing of control drum	3. Deflection due to thermal gradients	3. Analysis techniques improved, drum longitudinal slots added
NRX-A5	1. Support blocks peripheral and axial cracks	1. Stresses due to thermal gradients and mechanical loading	1. Serious problem at end of reporting period
	2. Liner tube flaring	2. Cluster components overheating caused liner tube expansion which coupled with tapered sleeve restraint caused flaring	2. Reduced liner tube length
	3. 1/3 of fuel elements broken	3. Reduced strength due to excessive corrosion plus inflow gradients and end of life	3. Increase emphasis on coating technology
	4. Hot buffer filler strip extensive breakage	4. Stresses due to thermal gradients	4. Eliminate hot buffer
NRX-A6	1. Beryllium reflector ring cracked	1. Stresses due to thermal gradients	1. Components tests verifying 3-D finite element analysis technique
	2. Support blocks cracks	2. Stresses due to thermal gradients	2. A predictable problem requiring extended efforts for similar components in the flight engine era of NERVA
	3. Cracking of peripheral composite cups and one tungsten cup	3. Stresses due to high non-symmetric transverse temperature gradient	3. Non-expected, protection cup performance not impaired

NERVA Engine Development

NERVA engine development was keyed to maximizing overall engine specific impulse, meeting various defined thrust levels, minimizing mass, and increasing longevity from an initial goal of one hour to ten hours. The flight engine design placed a high premium on reliability with a goal of 0.997 at a 90% confidence level.

For better engine performance, in addition to increasing the reactor fuel operating temperatures, continuous improvements were made in the reactor design configuration. Early reactors used tie-rods attached to the cold end core support plate that were cooled by hydrogen discharged into the nozzle chamber. Because the tie-rod coolant was at a fairly low temperature relative to the hydrogen passing through the main fuel elements, this lowered the rocket specific impulse. To avoid this reduction in specific impulse, later reactors and in the engine systems designed for space flight, tie-tubes were substituted for tie-rods. The tie-tubes were arranged as regeneratively cooled heat exchangers, with the coolant discharging into the core inlet. Also in the early reactors, a high core peripheral flow rate was used to protect the outer section of the core. This flow was steadily decreased to an almost negligible amount in order to increase the temperature of the hydrogen discharged into the nozzle chamber. A final optimization involved the use of a regenerative-heat exchanger-cooled peripheral system.

The engine cycle was also changed to increase the specific impulse. In the XE experimental engine, the hot bleed cycle was used (see Fig. 12.). In this cycle the turbopump was powered using some heated hydrogen extracted from the exit chamber and mixed with fluid from the reflector outlet. Because of the reduced pressure, the fluid could not be reintroduced into the main flow stream. Consequently, the turbine discharge was exhausted to space at a relatively low temperature when compared to the main nozzle chamber conditions. This significantly reduced the overall specific impulse of the engine.

In the evolution of the engine to a full flow, or "topping cycle," the turbine used fluid from the tie-tube outlet and the turbine working fluid was then discharged into the core inlet. This made it possible to significantly raise the specific impulse of the system. However, the turbine inlet temperature was now much lower than in the hot-bleed cycle. To compensate for the lower turbine temperature, a much higher turbine flow rate was needed. This led to the need for a much higher pump discharge pressure in the "topping cycle".

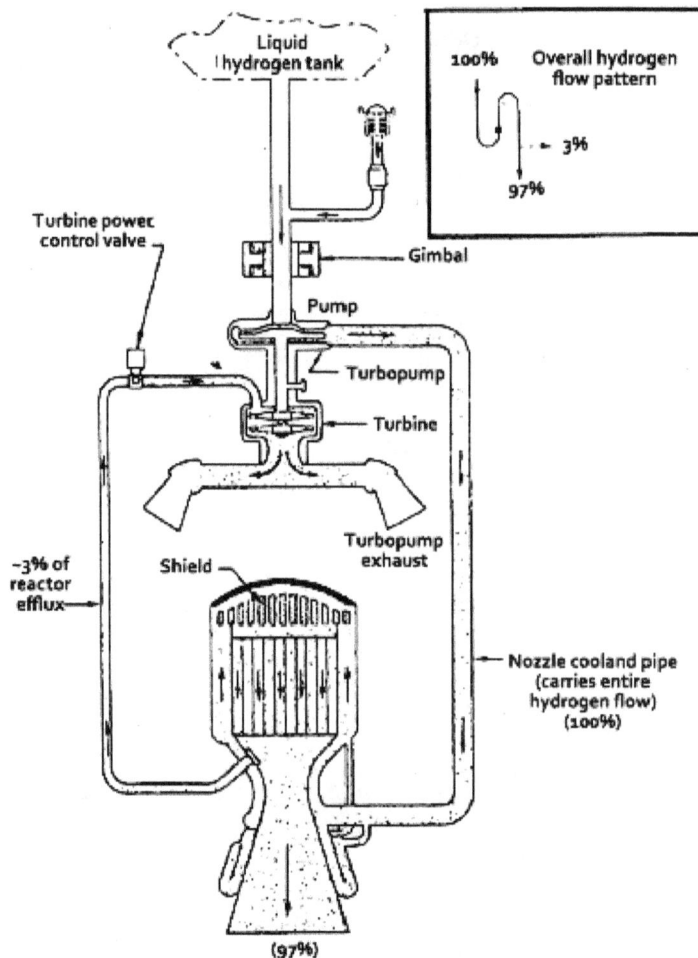

Fig. 12. A "hot bleed cycle" nuclear rocket reactor engine in which a small fraction of the reactor power is used to drive the propellant turbopump. *Courtesy of Department of Energy.*

The reliability requirement of 0.997 at 90% confidence level was a significant design driver in the flight version of the NERVA engine. To meet this requirement a methodology using probability in the design process was formulated. This process involved:

-   assigning probabilities and confidence levels to each part and component that must be meet to satisfy the overall engine design requirement;
-   identifying all possible failure modes in each part and component;
-   identifying all operational environments, both internal to the engine and external;
-   evaluating all operational modes including standby, startup, full power, and shutdown; and
-   evaluating synergistic failures; and developing the failure probabilities for each part and component.

This identified in many cases the need for additional information and testing to meet the assigned goals. And, in some case like in the valves, controls and instrumentation, the need for adding redundancy to the engine design. The use of probability in the design process was a major advancement in the engineering design process.

There was an evolution in engine designs throughout the NERVA program. Earlier designs, like XE', used the hot bleed cycle and fuel that produced lower chamber temperatures. Later designs used the "topping cycle" and with advances in fuel, projected higher chamber temperatures. Table 4 lists parameters for various Rover/NERVA engine designs. The experimental XE' was successfully tested. The NERVA engine was being designed for flight at the 337 kilonewton (kN) (75,000 lb$_r$) thrust level; while the small engine was being designed for a 72 kilonewton (16,000 lb$_r$) thrust level. The XE' engine would have had a flight specific impulse of about 6,960 m/s (710 s); the NERVA flight engine could have improved this to 8,085 m/s (825 s); and the small engine could have achieved a specific impulse of 8,575 m/s (875 s) had these nuclear rocket engine programs been conducted through the flight stage. The power levels for the various engine systems were: 1,140 MW for XE', 1,556 MW for NERVA, and 367 MW for the small engine. The higher specific impulse levels are reflected in the chamber temperatures that went from 2,270 K (demonstrated in the XE') to 2,695 K for the small engine.

Table 4 Rover Program engine characteristics

	XE'	NERVA	Small Engine
Thrust (kN)	245	337	72
Specific impulse (m/s)	6,970	8,085	8,575
Thermal power (MW)	1,140	1,566	367
Turbopump power (MW)	5.1		0.9
Turbopump speed (rpm)	22,270	23,920	46,950
Pump discharge pressure (MPa)	6.80	9.36	6.03
Engine flow rate (kg/s)	35.9	41.9	8.5
Chamber temperature (K)	2,270	2,360	2,695
Chamber pressure (MPa)	3.86	3.10	3.10

In effect, the small engine (see Fig. 13)[9] represented an accumulation of all of the knowledge gained in the nuclear rocket development program. Elements included: hydrogen as the propellant; the full-flow engine topping cycle; a single stage centrifugal pump with a single-stage turbine; a regeneratively cooled nozzle; and tie-tube core support elements. A radiation shield of borated zirconium hydride was incorporated above the reactor. This was used mainly to reduce neutron heating of the hydrogen propellant tank situated above the engine. It would also have provided some shielding for the payload and astronaut crew. Reactor control utilized six control drum drums and actuators with the control drums being located in the beryllium reflector. The small engine had five fluid control valves and actuators. The valving included a propellant tank shutoff valve (PSOV), a nozzle control valve (NCV), a turbine series control valve (TSCV), a turbine bypass control valve (TBCV), and a cooldown control valve (CCV). The PSOV was located at the bottom of the propellant tank and provided a tight seal against propellant leakage when the engine was not in use. The NCV adjusted the flow split between the nozzle coolant tubes and the tie-tubes. The TSCV was used to isolate the turbine during preconditioning and cooldown. The TBCV regulated the amount of the flow to the turbine and, consequently, the turbopump speed and flow rate. Finally, the CCV was used to regulate hydrogen flow for decay heat removal following engine operation. It was also used in conjunction with a small pump to provide pre-pressurization for the tank.

The small engine (pictured in Fig. 14) incorporated a nozzle with an area ratio of 100:1 (exit area to throat area). Part of it, out to an area ratio of 25:1, was regeneratively cooled. The 25:1 nozzle was followed by an uncooled skirt section that extended the overall nozzle out to the area ratio of 100:1. The uncooled nozzle skirt was hinged to facilitate packaging in the launch vehicle--a design arrangement that also provided room for a larger propellant tank in the spacecraft. The overall engine length was 3.125 meters with the nozzle skirt folded or 4.415 meters with the nozzle skirt in place. Also on Fig. 14 is a summary of various flow rate, pressure, and temperature points throughout the engine system. The chamber temperature was about 2,700 K.

Small engine component masses are listed in Table 5. The overall mass was 2,550 kg with the reactor (minus the shield) representing almost 1,600 kg of this amount and the turbopump only some 41 kg.

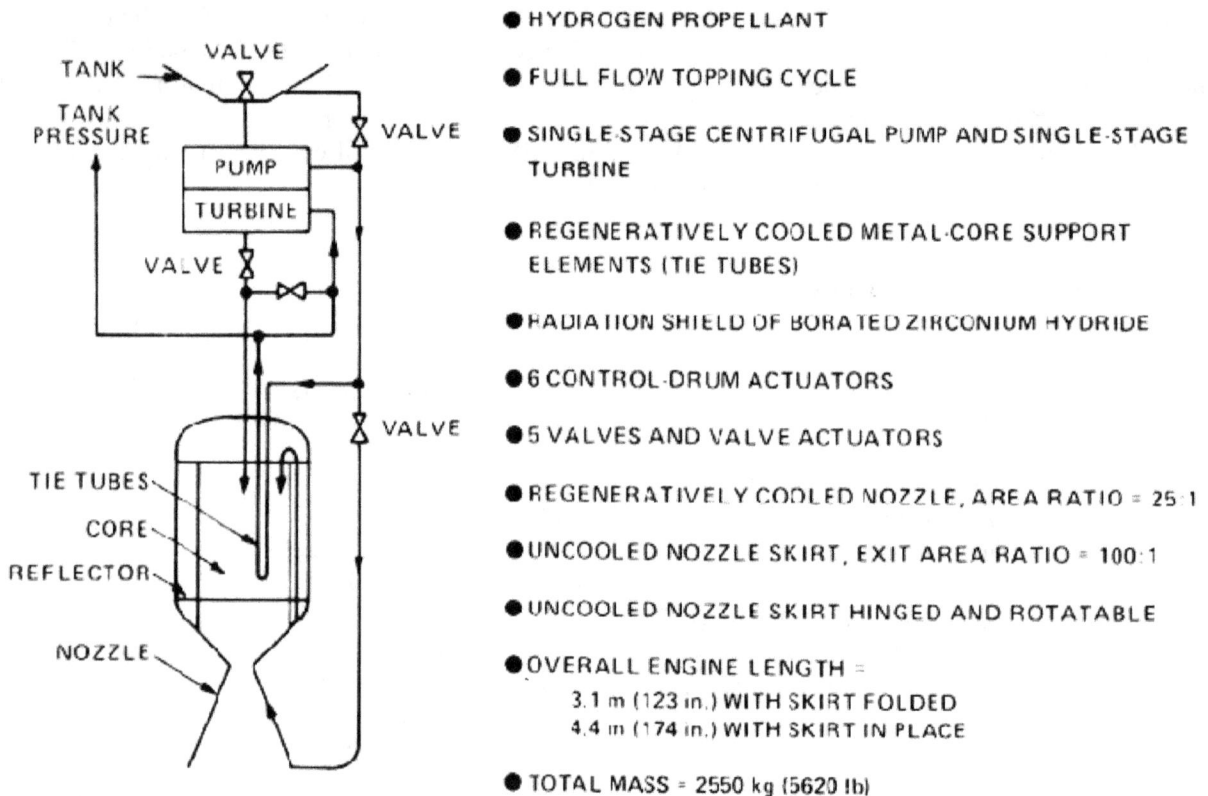

- HYDROGEN PROPELLANT
- FULL FLOW TOPPING CYCLE
- SINGLE-STAGE CENTRIFUGAL PUMP AND SINGLE-STAGE TURBINE
- REGENERATIVELY COOLED METAL-CORE SUPPORT ELEMENTS (TIE TUBES)
- RADIATION SHIELD OF BORATED ZIRCONIUM HYDRIDE
- 6 CONTROL-DRUM ACTUATORS
- 5 VALVES AND VALVE ACTUATORS
- REGENERATIVELY COOLED NOZZLE, AREA RATIO = 25:1
- UNCOOLED NOZZLE SKIRT, EXIT AREA RATIO = 100:1
- UNCOOLED NOZZLE SKIRT HINGED AND ROTATABLE
- OVERALL ENGINE LENGTH =
    3.1 m (123 in.) WITH SKIRT FOLDED
    4.4 m (174 in.) WITH SKIRT IN PLACE
- TOTAL MASS = 2550 kg (5620 lb)

Fig. 13. Schematic of Small Engine Nuclear Rocket design with a thrust of 71,714 N (16,125 lb$_f$).
After F. D. Durham, LA-5044-MS, Los Alamos National Laboratory.

The small engine reactor core (pictured in Fig. 15) was designed to produce 365 MW. It contained 564 hexagonally shaped (UC-ZrC)C composite fuel elements. Each fuel element had 19 coolant channels. There were 241 support elements, containing zirconium hydride (ZrH) as a neutron moderator. The core periphery included an outer insulator layer, a cooled inboard slat section, a metal wrapper, a cooled outboard slat section, and an expansion gap. The core was supported by an aluminum alloy plate on the cold end with the support plate resting on the reflector system. The reactor was contained in an aluminum pressure vessel. A beryllium barrel containing 12 control drums surrounded the core. The small engine reactor was designed for 83 degrees Kelvin per second (K / s) temperature transients.

Small Engine Design Conditions

Chamber Pressure	310 N/cm²
Chamber Flow Rate	8.51 kg/s
Chamber Temperature	2696 K
Reactor Power	367 MW
Specific Impulse	8580 m/s
Thrust	7297 N
Nozzle Flow Fraction	44.9 %
Turbine Bypass Flow Fraction	11.8 %
Turbopump Speed	4917 rad/s
Turbopump Shaft Work	0.93 MW
Pump Efficiency	65%
Turbine Efficiency	80 %
Nozzle Valve Area	2.63 cm²
Turbine Control Valve Area	3.02 cm²

State Point Description	Flow Rate	Pressure	Temperature
Pump Inlet	8.51	0.12	17.0
Pump Exit	8.51	6.03	19.8
Tie Tube Manifold Inlet	4.05	5.72	20.3
Tie Tube First Pass Exit	4.05	5.38	56.9
Tie Tube Exit	4.05	5.02	428.9
Slat Manifold Inlet	0.64	5.72	20.3
Slat First Pass Exit	0.64	5.24	167.1
Slat Exit	0.64	5.02	431.5
Turbine Inlet	4.13	4.86	428.6
Turbine Exit Mixed	4.69	4.13	415.6
Turbine Bypass Inlet	0.55	4.86	428.6
Nozzle Inlet	3.83	4.63	240.4
Nozzle Exit	3.83	4.21	21.4
Reflector Exit	3.83	4.21	294.9
Shield Inlet	8.51	4.06	361.0
Core Inlet	8.51	3.96	370.1
Fuel Element Exit	8.33	3.1	2728.0
Core Bypass Exit	0.18	3.1	370.1
Chamber	8.51	3.1	2695.8

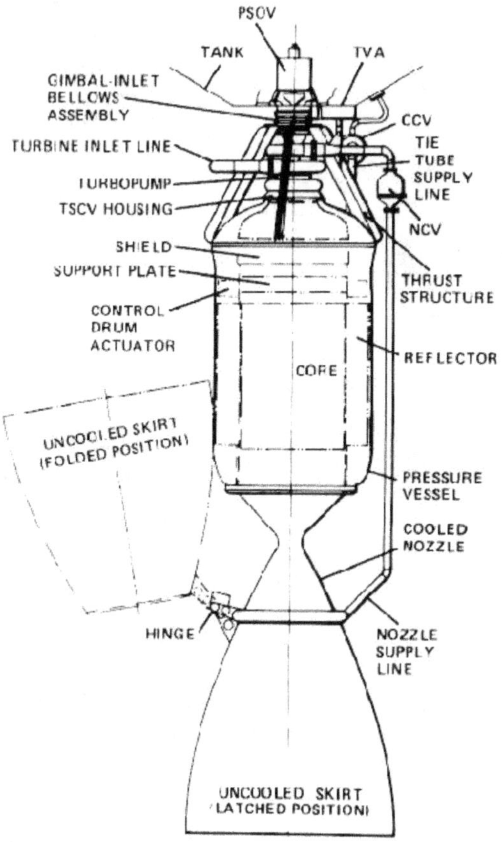

Fig. 14. Small Engine Nuclear Rocket design conditions. *After F. D. Durham, LA-5044 MS, Los Alamos National Laboratory.*

Table 5. Small Engine Nuclear Rocket mass estimates (kg).

Reactor core and hardware	868
Reflector and hardware	569
Shield	239
Pressure vessel	150
Turbopump	41
Nozzle and skirt assembly	224
Propellant lines	15
Thrust structure and gimbal	28
Valves and actuators	207
Instrumentation and electronics	159
Contingency	50
Total	2,550

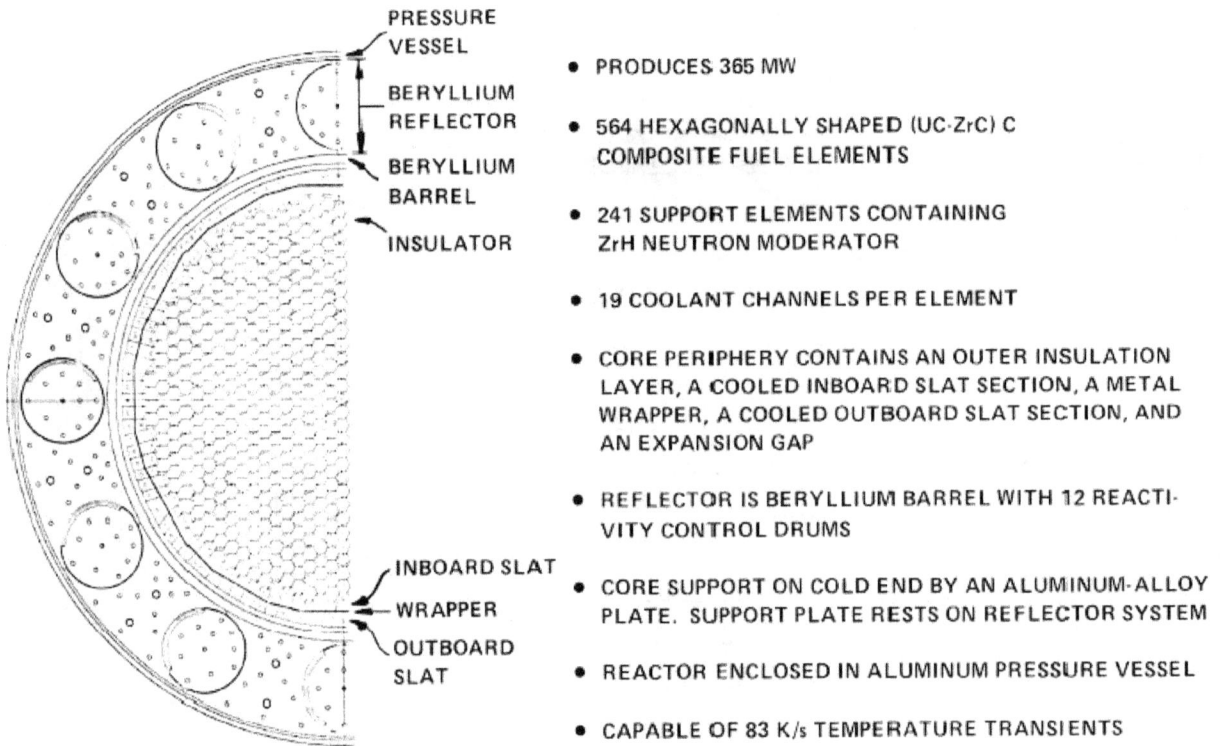

PRESSURE VESSEL
BERYLLIUM REFLECTOR
BERYLLIUM BARREL
INSULATOR
INBOARD SLAT
WRAPPER
OUTBOARD SLAT

- PRODUCES 365 MW
- 564 HEXAGONALLY SHAPED (UC-ZrC) C COMPOSITE FUEL ELEMENTS
- 241 SUPPORT ELEMENTS CONTAINING ZrH NEUTRON MODERATOR
- 19 COOLANT CHANNELS PER ELEMENT
- CORE PERIPHERY CONTAINS AN OUTER INSULATION LAYER, A COOLED INBOARD SLAT SECTION, A METAL WRAPPER, A COOLED OUTBOARD SLAT SECTION, AND AN EXPANSION GAP
- REFLECTOR IS BERYLLIUM BARREL WITH 12 REACTIVITY CONTROL DRUMS
- CORE SUPPORT ON COLD END BY AN ALUMINUM-ALLOY PLATE. SUPPORT PLATE RESTS ON REFLECTOR SYSTEM
- REACTOR ENCLOSED IN ALUMINUM PRESSURE VESSEL
- CAPABLE OF 83 K/s TEMPERATURE TRANSIENTS

Fig. 15. Reactor core for Small Engine Nuclear Rocket. *After F. D. Durham, LA-5044 MS, Los Alamos National Laboratory.*

Additional details about the small engine fuel modules are shown in Fig. 16. The fuel provided both the energy for heating the hydrogen and the heat transfer surface to accomplish propellant heating. It was enriched (93.15%) uranium-235 in a composite matrix of UC-ZrC solid solution and carbon. The flow channels were coated with zirconium carbide to protect against possible hydrogen reactions. The tie-tubes were used to transmitted the core axial pressure load from the hot end of the fuel elements to the core support plate. They also acted as the energy source for the turbopump and contained and cooled the ZrC moderator sleeves. The tie-tubes consisted of a counterflow heat exchanger made of Inconel 719 and a zirconium moderator with zirconium carbide insulation sleeves. Their overall length was approximately 0.9 meter.

Extensive development of the engine components was accomplished during the NERVA program. This development represents the state-of-the-art of nuclear rocket non-nuclear components.

The turbopump provides the pressure to drive the hydrogen through the engine. The hydrogen is stored in a liquid state in the propellant tank. Turbopump development is summarized in Fig. 17. The XE' turbopump was a single-stage, radial exit flow, centrifugal pump with aluminum impeller, a power transmission that coupled the pump to the turbine, and a two stage turbine with stainless steel rotors. Operating experience on the NRX/EST engine, this pump performed eight starts and operated 54.4 minutes at high power. Additional operational experience was gained In the XE' engine where the turbopump performed 28 starts and restarts, including runs to rated power. One difficulty encountered in the XE' tests was that the shaft system bound at the bearing. The solution was to increase clearance and to improve alignment. Bearings, in fact, probably represent one of the few life-limiting components among the non-nuclear components of a nuclear rocket system. During the NERVA Program, the solution to the bearing problem appeared dependent upon maintaining adequate cooling to reduce wear.

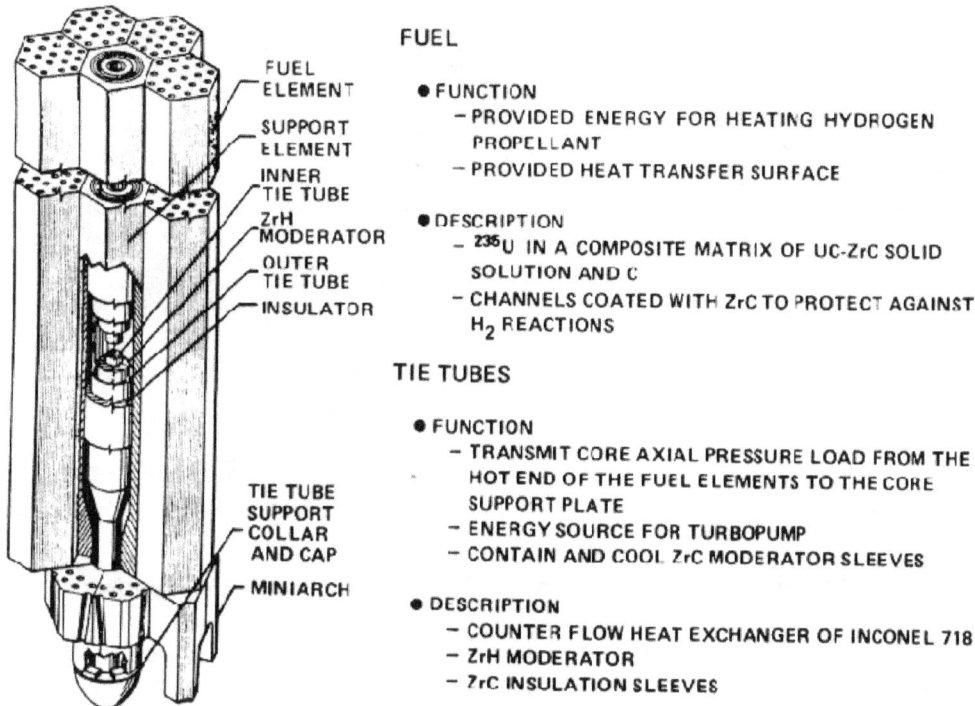

FUEL ELEMENT
SUPPORT ELEMENT
INNER TIE TUBE
ZrH MODERATOR
OUTER TIE TUBE
INSULATOR
TIE TUBE SUPPORT COLLAR AND CAP
MINIARCH

FUEL
● FUNCTION
  – PROVIDED ENERGY FOR HEATING HYDROGEN PROPELLANT
  – PROVIDED HEAT TRANSFER SURFACE
● DESCRIPTION
  – ^{235}U IN A COMPOSITE MATRIX OF UC-ZrC SOLID SOLUTION AND C
  – CHANNELS COATED WITH ZrC TO PROTECT AGAINST H_2 REACTIONS

TIE TUBES
● FUNCTION
  – TRANSMIT CORE AXIAL PRESSURE LOAD FROM THE HOT END OF THE FUEL ELEMENTS TO THE CORE SUPPORT PLATE
  – ENERGY SOURCE FOR TURBOPUMP
  – CONTAIN AND COOL ZrC MODERATOR SLEEVES
● DESCRIPTION
  – COUNTER FLOW HEAT EXCHANGER OF INCONEL 718
  – ZrH MODERATOR
  – ZrC INSULATION SLEEVES

Fig. 16. Fuel module for small engine nuclear rocket. *After F. D. Durham, LA-5044-MS, Los Alamos National Laboratory.*

- FUNCTION
  - PRESSURIZE THE PROPELLANT FOR THE ENGINE FEED SYSTEM

- DESIGN CONDITIONS

	XE'	NERVA	SMALL ENGINE
PUMP DISCHARGE PRESSURE (MPa)	6.69	9.36	6.03
PUMP FLOW RATE (kg/s)	35.8	20.9 – 41.7	8.5
TURBINE TEMP (K)	648	154	429
TURBINE FLOW RATE (kg/s)	3.32	19	4.1
SHAFT SPEED (rpm)	21989	23920	46952

- CONSTRUCTION (XE)
  - SINGLE-STAGE, RADIAL-EXIT-FLOW-CENTRIFUGAL PUMP WITH AN ALUMINUM IMPELLER, A POWER TRANSMISSION THAT COUPLES THE PUMP TO THE TURBINE, AND A TWO-STAGE TURBINE WITH STAINLESS STEEL ROTORS

- REACTOR TESTS EXPERIENCE
  - NRX/EST 8 STARTS INCLUDING 54.4 MINUTES AT HIGH POWER
  - XE-28 STARTS/RESTARTS INCLUDING RUNS TO RATED POWER

- POTENTIAL PROBLEMS
  - SHAFT SYSTEM BINDING AT BEARING COOLANT LABYRINTH IN XE TESTS (INCREASE CLEARANCE AND IMPROVE ALIGNMENT)
  - BEARING LIFE

Fig. 17. Turbopump technology. *Courtesy of Aerojet General Corporation.*

The nozzle assembly expands the heated gas as it emerged from the reactor in order to maximize the thrust. A typical nozzle assembly is shown in Fig. 18. The design conditions for the NERVA flight engine included a thrust level of 337 kN using a nozzle with an area ratio of 100:1, a service life of 10 hours, a reliability level of less than 4 failures in 10^4 flights, a chamber pressure of 3.1 MPa, a chamber temperature of 2,360 K, a propellant flow rate of 41.6 kg/s, and operation of the coolant channels at temperatures between 28 and 33 K. In the regeneratively cooled portion of the nozzle, an aluminum alloy jacket with stainless steel for the coolant channels was used. The nozzle extension to an area ratio of 100:1 was made of graphite; this portion of the nozzle assembly did not require cooling. The major unresolved engineering problems in achieving a 10 hour operating life was resolving possible stress problems in the aluminum alloy and fabrication difficulties. However, the use of ARMCO 22-13-5 appears to resolve the stress problems and fabrication difficulties also appear to have been overcome in the program.

- FUNCTION
  - EXPAND GAS TO PROVIDE MAXIMUM THRUST

- DESIGN CONDITIONS (NERVA ENGINE)
  - THRUST 334 kN
  - AREA RATIO 100:1
  - SERVICE LIFE 10 h
  - RELIABILITY 0.9998
  - CHAMBER PRESSURE 3.1 MPa
  - CHAMBER TEMPERATURE 2360 K
  - FLOW RATE 41.6 kg/s
  - COOLANT CHANNEL 28–33 K

- CONSTRUCTION
  - LH_2 COOLED SECTION TO 24:1 OF ARMCO 22–13–5 JACKET AND CRES 347 COOLANT CHANNELS

  - GRAPHITE NOZZLE EXTENSION UNCOOLED TO 100:1
  - U-TUBE CONSTRUCTION IN DIVERGENT SECTION

- UNRESOLVED PROBLEMS
  - USE OF ARMCO 22–13–5 APPEARS TO HAVE RESOLVED STRESS PROBLEMS. FABRICATION PROBLEMS ALSO APPEAR RESOLVED

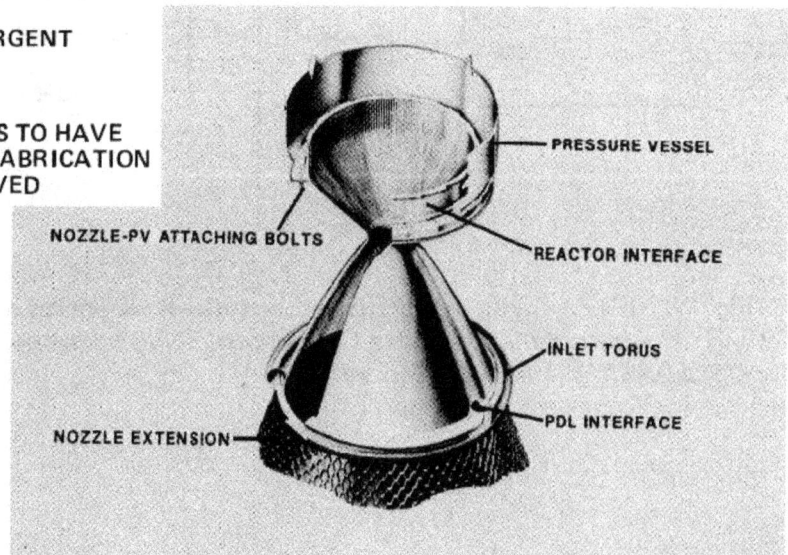

Fig. 18. Nozzle assembly. *Courtesy of Aerojet General Corporation.*

Hydrogen flow was controlled by a valve arrangement. These valves were binary valves, in-out control valves and check valves. The small engine design had just five valves. However, to increase system reliability in the design of the NERVA flight engine, two turbopumps (either one of which could provide full flow and pressure to the engine) were included. In addition, redundant valves were incorporated into the engine design. In order to provide switching between the turbopumps and to provide high reliability (to meet the 0.997 confidence level of 90%) by backing up each valve in case it should fail, some 26 valves were needed, as shown in Fig. 19. Table 6 lists NERVA valve and actuator characteristics. Reactor operating experience for the various valves was obtained in both the NRX/EST and XE' engine tests. The major potential problems appeared to be from seal damage by contaminants, erroneous position indicators, and leakage from lip-seal tolerances.

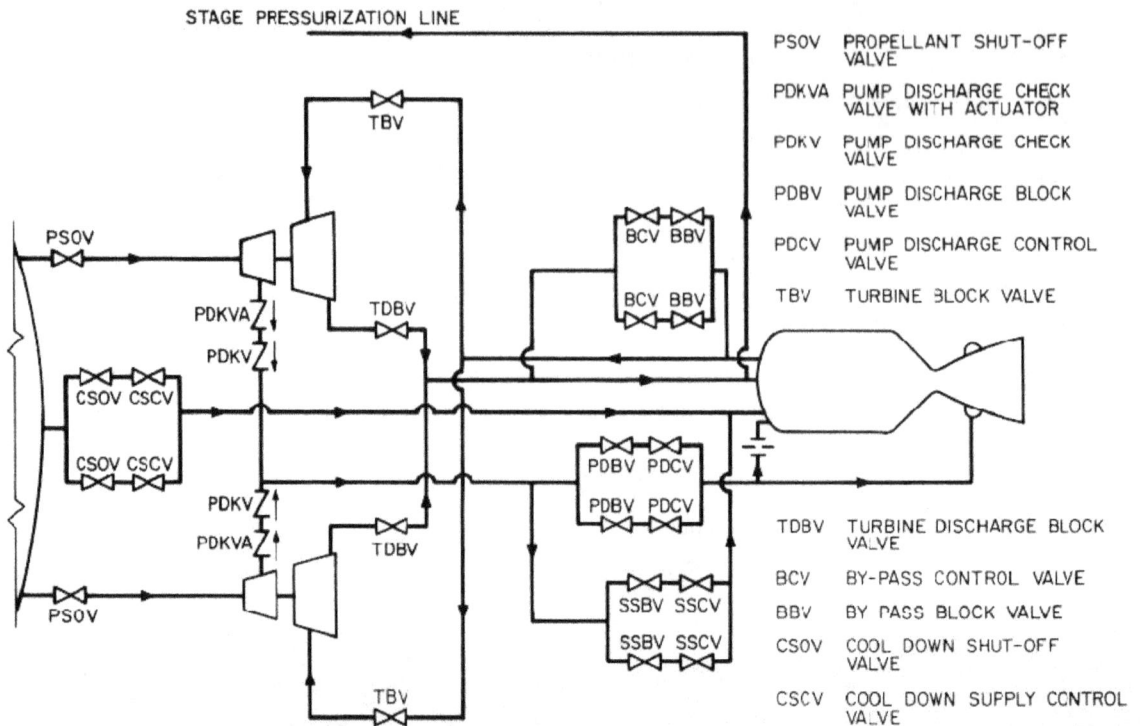

Fig. 19. NERVA engine schematic with parallel turbopumps and redundant engine valves. *After J. H. Altseimer et al., "Operating Characteristics and Requirements for the NERVA Flight Engine," J. Spacecraft, Vol. 8, No. 7, July 1971.*

Table 6. NERVA valve and actuator characteristics.

	Cooldown Control Valve (CCV)	Nozzle Control Valve (NCV)	Propellant Shutoff Valve (PSOV)	Turbine Bypass Control Valve (TBCV)	Turbine Shutoff Control Valve (TSCV)	Turbine Valve Actuator (TVA)	Control Drum Actuator (CDA)
Actuator size	small	small	large	large	large	large	small
Seat diameter, cm	2.54	2.54	15.24	3.175	11.43	—	—
Seat area, cm^2	5.06	5.06	182	7.93	103	—	—
Stroke	1.52 cm	1.52 cm	5.08 cm	1.52 cm	3.18 cm	$\pm$ 0.044 rad*	3.14 rad
Load max. velocity	0.15 cm/s^{-1}	0.15 cm/s^{-1}	0.5 cm/s^{-1}	2.5 cm/s^{-1}	0.2 cm/s^{-1}	0.007 rad/s^{-1}	0.2 rad/s
Seating force, N	3,550	3,558	13,340	2,800	10,000	—	—
Plus max. ΔP at rated speed, N cm^{-2}	760	275	20.7	103	103	—	—
Lead screw diameter, cm	1.27	1.27	1.9	1.9	1.9	—	—
Lead screw advance, cm rad^{-1}	0.0809	0.0809	0.0809	0.0809	0.0809	—	—
Max stepping rate, steps s^{-1}	62	62	200	210	75	143	122
Motor gear ratio	50	50	50	10	50	380	917
Torque at speed $N - m$	—	—	—	—	—	3,250	34

*rad = radian
CCV and NCV are identical
PSOV and TCCV have same actuator

The pressure vessel and enclosure was to support the components of the reactor assembly, to form a pressure shell for the hydrogen propellant, and to transmit thrust to the thrust structure. This is pictured in Fig 20. NERVA engine design conditions required a maximum flow rate of 37.6 kg / s, a maximum pressure of 8.66 MPa, temperatures ranging from 20 to 180 K, a reliability of less than three failures in 10^6 flights, and a service life of 10 hours. Designs similar to the flight engine were used in the five NRX tests and the XE' engine. The pressure vessel consisted of a cylinder incorporating top closure with bolts and seals. A one-piece extruded forging of aluminum alloy 7075-773 was used with a surface coating of Al_2O_3. The major engineering features still under development at the time of program termination included optimum ways of assuring bulk preload and methods of finalizing the surface coatings.

- FUNCTION
  - SUPPORTS COMPONENTS OF REACTOR ASSEMBLY TO FORM A PRESSURE SHELL FOR THE HYDROGEN PROPELLANT
  - TRANSMITS THRUST TO THE THRUST STRUCTURE

- DESIGN CONDITIONS (NERVA ENGINE)
  - MAXIMUM FLOW RATE 37.6 kg/s
  - MAXIMUM PRESSURE 8.66 MPa
  - TEMPERATURE RANGE 20 – 180 K
  - RELIABILITY 0.999997
  - SERVICE LIFE 10 h

- REACTOR TESTS
  - NRX–5 TESTS, XE'

- CONSTRUCTION
  - CYLINDER, TOP CLOSURE, BOLTS AND SEALS
  - ONE-PIECE EXTRUDED FORGING OF ALUMINUM ALLOY 7075–773
  - SURFACE COATING Al_2O_3

- UNRESOLVED DESIGN ITEMS ON NERVA ENGINE
  - METHOD TO ASSUME BOLT PRELOAD
  - FINALIZE SURFACE COATINGS

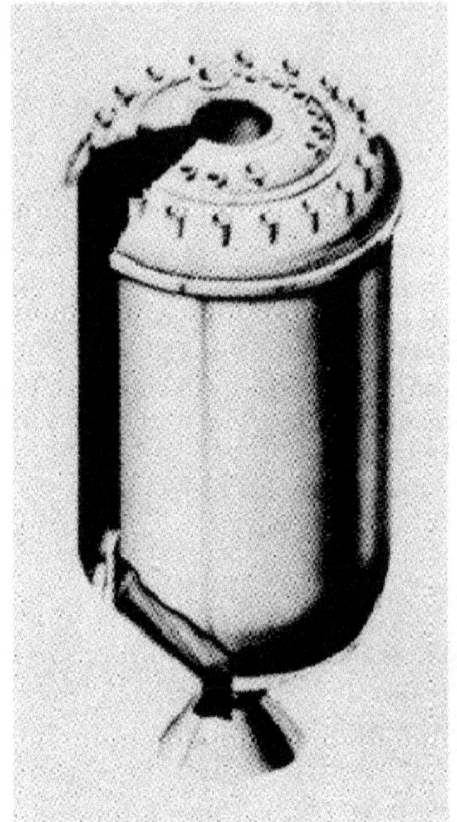

Fig. 20. Pressure vessel and enclosure. *Courtesy of Aerojet General Corporation.*

Controls and instrumentation underwent extensive development during the NERVA Program. Actuators for the control drums were pneumatic. These actuators were used on the XE' engine without apparent degradation or anomalies. Thermocouples demonstrated performance at 2,667 K for one hour without degradation. In addition, displacement, pressure, and vibration sensors were developed for several hours of operation. However, instrumentation with an accuracy of I percent or better required further development.

Extensive efforts were spent on the understanding how to startup the engine. The startup characteristics of a nuclear rocket engine are shown in Fig. 21. The engine components must be conditioned before high power can be reached. The turbopump, nozzle, reflector, and core inlet were all designed to operate at low temperatures. When the pump shutoff valve was first opened, the pump tended to vaporize the working fluid until sufficient fluid passed through the device, chilling it down to cryogenic conditions. The nozzle also tended to be a fluid choke point, as were the core and reflector inlets. Therefore, a certain amount of hydrogen must be passed through the system to remove thermal energy stored in the line, valves, and reflector. Once this chilling down process has been accomplished, the pump could be started and would operate normally. It was determined that approximately one minute of fluid flow was necessary to accomplish this conditioning. While the flow components are being conditioned, the reactor can be brought up to a low power level. This was accomplished by programming the reactor drums out rapidly almost to the cold critical point, and then using a slow transient program. After an appreciable temperature was sensed in the chamber, the reactor could be switched over to closed-loop temperature control. This scheme did not require any neutronic instrumentation. When appreciable power had been achieved and the turbopump running, engine heating could then be accelerated at the rate of 83 K / s. NRX/EST and XE engine experience indicate that the nuclear rocket engine can be controlled in a predictable and safe manner. The XE engine was operated safely and as predicted over the wide range of operating conditions in several control modes including the range encompassed by the solid lines of Fig. 22.[10]

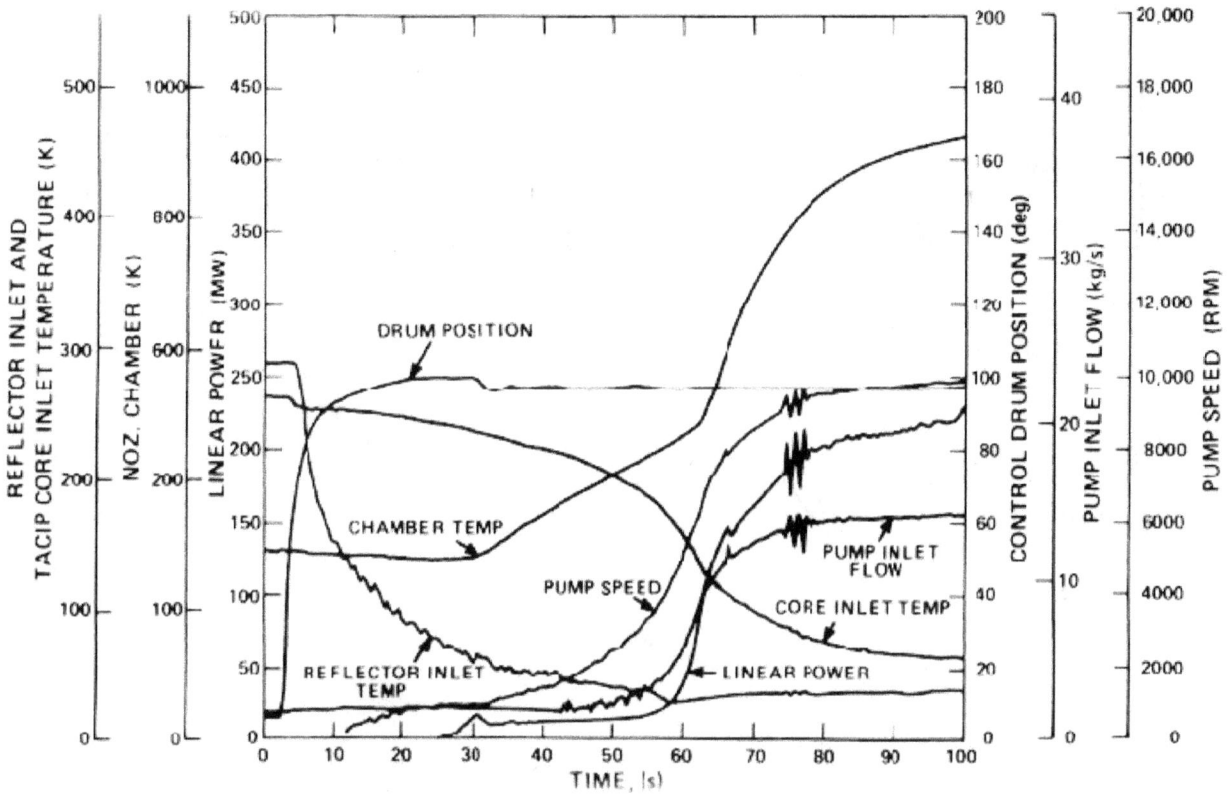

Fig. 21. Startup characteristics of a nuclear rocket

One unique characteristic of a nuclear thermal rocket is the ability to throttle the engine system. Throttling a flight engine system provides a means of reducing thrust towards the termination of a burn in order to obtain a final trim on velocity, a means to minimize the decay heat removal cooling requirements, or to provide a safe retreat position in case of a possible component malfunction. Decay heat occurs in the reactor as a result of radioisotope elements from the fissioning of uranium. The fission products elements have many half lives so that the radioactivity persists for long periods of time. Fig. 23 shows the decay heat for a 60 minute operating time of a 1126 MW reactor. The after heat removal is minimized by going to a lower thrust, with its corresponding lower power and reduced neutronics decay heat. This throttling maneuver can be performed with no loss in engine specific impulse.[11]

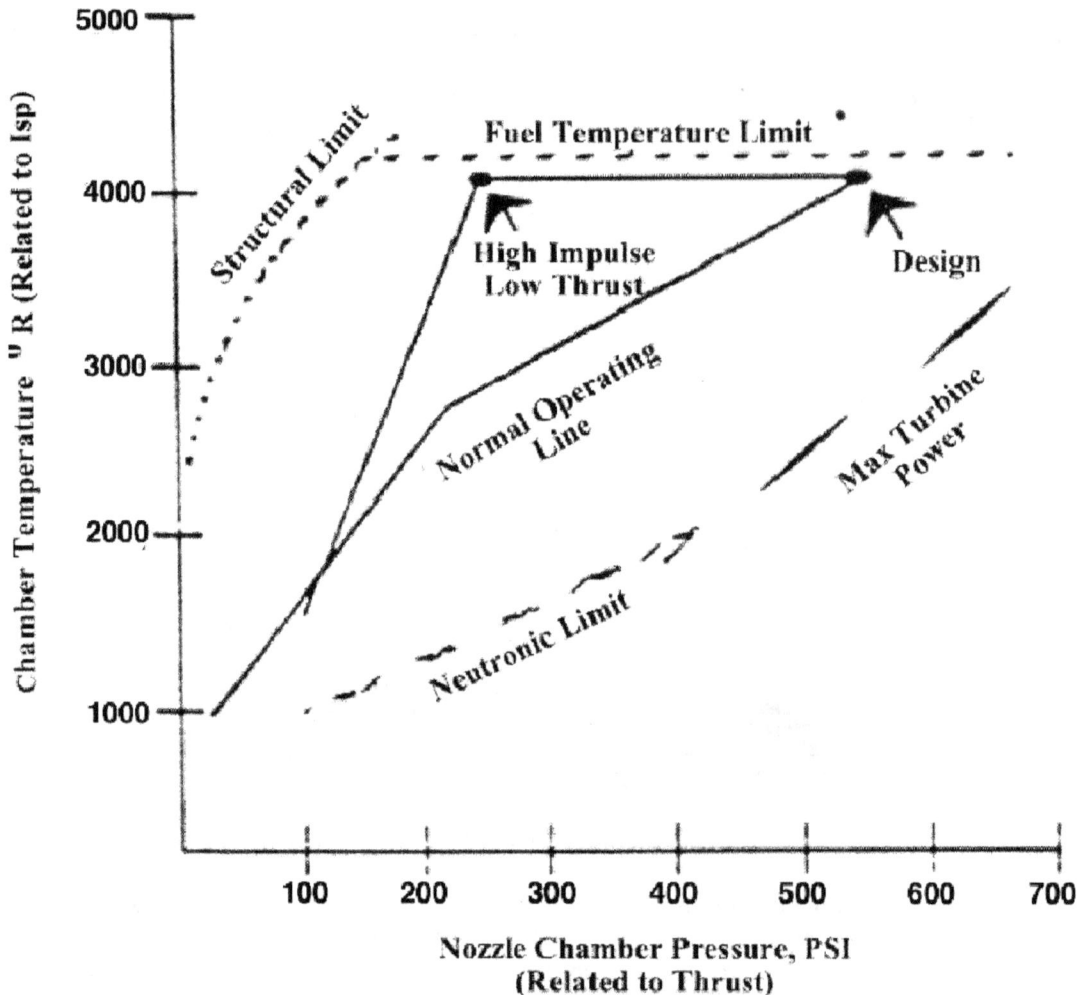

Fig. 22. Engine operating map.

Shutting down a nuclear rocket results in probably the most unique part of a nuclear rocket engine operational cycle. The initial phase is a retreat from power until the reactor control drums are fully turned to the minimum reactivity position. Though the reactor is subcritical, heat continues to be generated by the radioactive decay products. This decay heat continues at a level that requires coolant based on prior operating history. The two phases of shutting down are: (1) shutdown which is the retreat from power to the point where the turbopump is no longer required to cool the reactor, and (2) afterheat removal stage where tank pressure is sufficient to cool the reactor.

The afterheat should be removed in such a way as to create a minimum penalty to engine performance. When the power levels are sufficiently low, the engine itself will radiate the energy to space. However, most of the decay heat that is generated at fairly high power, shortly after engine shutdown, must be removed in another way. The method is to use tank propellant to cool the reactor at as high a temperature as possible. The high temperature allows efficient use of the propellant. The thrust which is generated contributes to the total velocity increase that the vehicle requires for its mission in space. The afterheat is removed by pulsing the flow through the engine. At the high end of a selected temperature range, such as 722 K (1300 R), the pulse of flow is initiated. When the reactor is cooled to a lower temperature, such as 667 K (1200 R), the pulse of flow is terminated. Approximately 3 - 4% of the total engine impulse is generated during the cooldown afterheat removal phase.

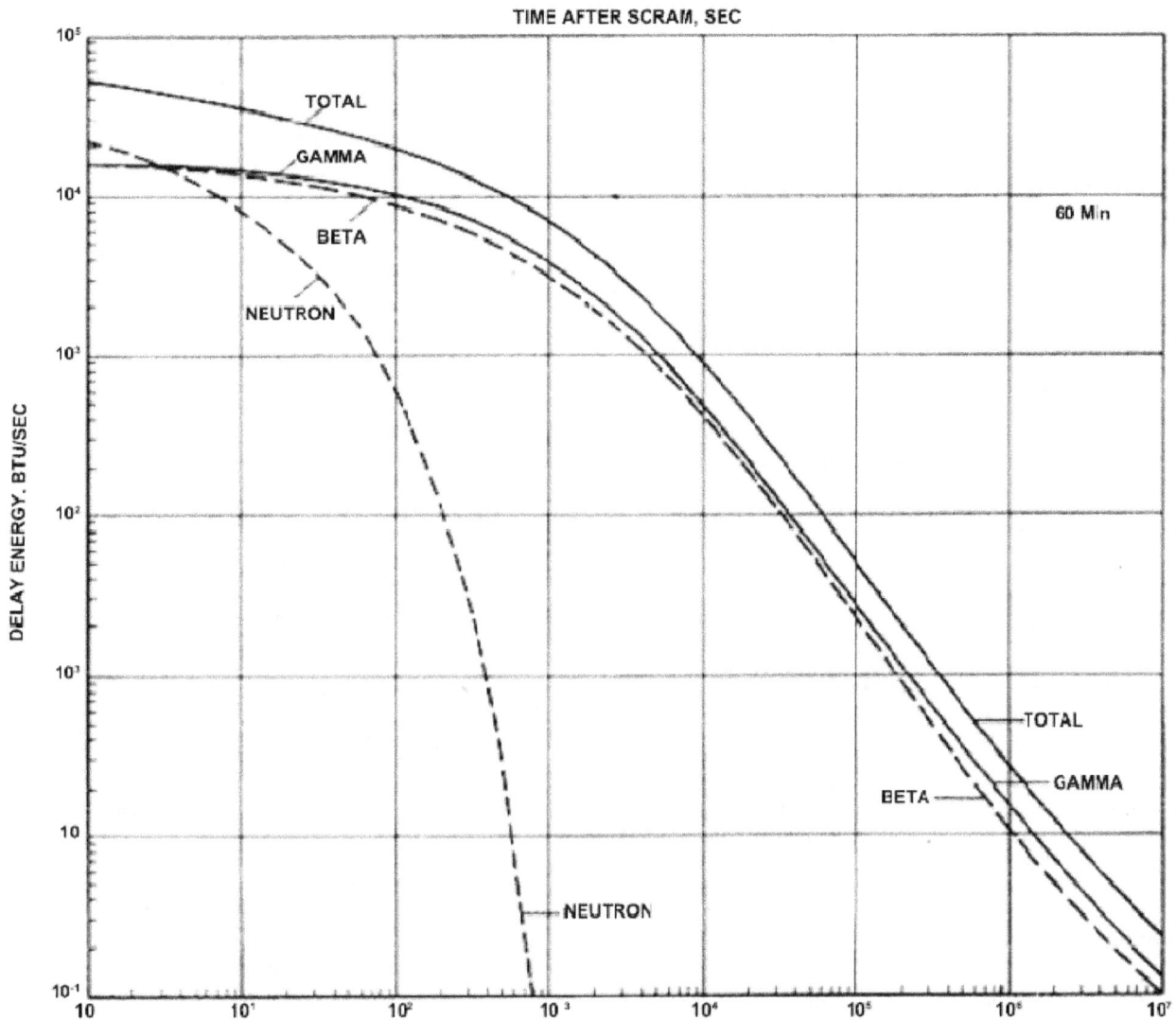

Fig. 23. Decay heat for 60 minute operating time.

NERVA Safety Program[12]

The NERVA program provides an excellent data base for developing NTP engines. The value of much of the work was not the numbers generated, but the forced attention to design details and the recognition of uncertainty. It forced designers to consider every aspect of safety from the initiation of the design process. Table 4 summarizes the NERVA safety plan. All of the potential flight failure accidents were examined in the NERVA program, and design and operational solutions developed.

Accidental insertions of reactivity could occur from either: (1) a control system malfunction, (2) water flooding, or (3) core compaction on impact. The energy release, if an accident supercritical condition occurred, depends upon (1) the amount of reactivity inserted, (2) the rate of insertion, (3) the initial state of the reactor (e.g., hot or cold), and (4) the quenching or shutdown mechanism. Rapid insertions of large amounts of reactivity would be accompanied by releases of kinetic energy, which physically disrupt the reactor. The KIWI-TNT test demonstrated the effects of large and rapid reactivity insertion. Special actuators were used to achieve the desired reactivity rates. The excursion released 10,000 MW(s) of energy and completely dismantled the core in a mechanical (not nuclear) explosion.

Table 4. NERVA Safety Requirements

- The means for preventing the inadvertent attainment of reactor criticality through any credible combination of failures, malfunctions, or operations during all ground, launch, flight, and space operations.

- A destruct system during launch and ascent to assure sufficient dispersion of the reactor fuel upon earth impact to prevent nuclear criticality with the fuel fully immersed in water.

-The means for preventing credible core vaporization or disintegration or violation of the thrust-load path to the payload.

- Diagnostic instrumentation adequate to detect the approach of a failure or an event that could injure the crew or damage the spacecraft and the provisions to preclude such an event.

- The capability for remote override of the engine programmer by the crew and ground control as well as for remote shutdown independent of the engine program

- An engine control system capability to preclude excessive or damaging deviations from programmed power and ramp rates.

- Provide an emergency mode on the order of 30,000 lb-thrust, 500 s specific impulse and 10^8 lb-sec total impulse.

The planned nuclear rocket engine stage was a modified Saturn vehicle, with the nuclear upper stage replacing the S-IVB. The potential energy releases of the booster propellants as a result of booster failure was a predominant factor in range safety. The Saturn booster fueled with liquid oxygen and RP-l included 2,180,000 kg (4,800,00 lb) of propellants and the S-II stage fueled with liquid oxygen and liquid hydrogen included 386,000 kg (850,000 lb) of propellants. In case of a destruct, it was calculated that 10 percent of the Saturn booster and 60 percent of the S-11 stage kinetic energy, or the equivalent of 218,000 kg (480,000 lb) of TNT from the former and 231,000 kg (510,000 lb) of TNT from the latter, needed to be considered in kinetic energy release. The nuclear stage included a destruct system that was integrated with the booster destruct system. In addition, an engine destruct system was tied to the nuclear stage destruct system. Therefore, if vehicle or nuclear stage destruct action was necessary, the reactor would also be safely destroyed. An ordinance destruct system would fragment the reactor into particles small enough to remain aloft as aerosols to be burned up upon reentry into the Earth' s atmosphere or with so little activity upon reaching the Earth's surface that they would not present a hazard.

The development of neutron poison systems to "safe" the reactor during its transport to the missile test site, during ground handling, and possibly during the early stages of launch was a primary thrust of the nuclear safety program. A redundant poison approach was pursued in which poisons could be inserted and reinserted into the core through the nozzle opening and reactor control elements could be locked. Therefore, if the control elements are inadvertently withdrawn, the core poisons could override the resultant reactivity insertion. Conversely, if the core poisons were withdrawn, the locked control system alone could safe the reactor.

A number of advanced countermeasures were also considered. Propulsion guidance interlocks were considered to interlock the propulsion and guidance systems in a manner to activate thrust termination in the event of guidance failures during orbital start-up or re-start to preclude prompt re-entry. Retro systems for inducing downrange impact in the event of late nuclear stage aborts during orbital injection to preclude random re-entry was another idea. Satellite interceptions might utilize ground-to-air or air-to-air missile systems to intercept and destroy nuclear rocket reactors or induce their impact into pre-determined marine disposal areas. Another idea considered was the use of auxiliary rockets to carry the nuclear rocket into orbit in case of late pre-orbital injection thrust failures or to transfer the nuclear stage to orbits of higher perigee in case of orbital start-up failures. This would provide additional decay time and also preclude prompt random re-entry. Automatic malfunction sensors and countermeasure initiators using on-board malfunction sensors in the nuclear stage (to detect guidance, thrust, or propellant malfunctions connected to automatic on-board initiators which execute destruct or countermeasure action, if necessary) were also being evaluated.

The NERVA Safety Plan established many requirements for flight safety.[13] It stated, for example, that a maximum effort was to be directed toward eliminating from the engine design those single failures or credible combinations of errors and failures which could endanger mission completion, the flight crew, the launch crew, or the general public. If this effort proved impossible or resulted in an excessive penalty, redundancies internal to the component in question were to be considered. If this alternate approach also proved ineffective, ways in which other components could compensate were to be investigated. Where no practical solutions were found in inherent design and where credible single or multiple failures could jeopardize crew or population safety, countermeasures or techniques such as maintainability and alternative operating modes were to be explored. Further, if the planned mission was to be abandoned because of an engine failure, provisions were to be made for engine operation in an emergency mode to affect safe crew return and to prevent danger to the Earth's population.

Operation in the emergency mode was to allow optimum use of remaining propellant commensurate with the failure and, at a minimum provide engine performance on the order of 30,000-thrust and 500-sec specific impulse. In addition, the engine was to be capable of delivering a minimum controllable total impulse of 10^8 lb-sec, including the impulse derived from the cooldown propellant. This total impulse was to be obtainable in a single thrust cycle with the powered-operation portion of the cycle at or above the specified thrust and specific-impulse minimums. This goal was to be obtainable from all operating phases of the engine operation, and provision was to be made for coolant up to five hours prior to entering the emergency mode. Final cooling was to preclude engine disintegration and - if possible at no addition risk to population, passengers, or crew - was to preserve the engine in a restartable condition.

Additional NERVA safety design requirements were to have the engine incorporate the following features:
- The means for preventing the inadvertent attainment of reactor criticality through any credible combination of failures, malfunctions, or operations during all ground, launch, flight, and space operations.
- A destruct system during launch and ascent to assure sufficient dispersion of the reactor fuel upon Earth impact to prevent nuclear criticality with the fuel fully immersed in water.
- The means for preventing credible core vaporization or disintegration or violation of the thrust-load path to the payload.
- Diagnostic instrumentation adequate to detect the approach of a failure or an event that could injure the crew or damage the spacecraft and the provisions to preclude such an event.
- The capability for remote override of the engine programmer by the crew and ground control as well as for remote shutdown independent of the engine program.
- An engine control system capability to preclude excessive or damaging deviations from programmed power and temperature ramp rates.

Because of these safety concerns and the often indistinguishable relationship between safety and reliability, the NERVA reliability program was a significant adjunct to the safety program.

Flight safety analysis was divided into three parts: malfunction analyses, fault tree analyses, and contingency analyses. Malfunction analyses were performed with a computer model and depict all the system effects of the failure of components. Fault tree analysis used the deductive process by which an undesirable event was postulated and possible malfunctions which caused the event were systematically analyzed. Contingency analyses addressed component failures and how they were detected, system consequences of the failures, contingency actions required, and the time in which the contingency action must be performed.[14]

Analysis of component failures indicates about 3 failure probabilities per 1000 engine cycles for catastrophic failures. (Analysis was only performed on the non-nuclear engine components, but a review of the nuclear subsystem led to this overall number.)[15]

Designers had primary responsibility to prove that a component met specifications. The technique chosen to ensure that the reliability goal would be met was Failure Mode Analysis. Failure Mode Analysis (FMA) is a systematic method used to ensure that components have high, inherent reliability. The FMA developed for NERVA clearly defined the conditions for success. A probability equation was written to express each condition. These equations were then used to define the principal distributions and to provide an indication of the kind of analysis performed.

A thorough, unbiased narrative listing all credible ways that failures can occur was written so that changes could be identified and used to eliminate those failures or minimize their effects. This listing gave insight into fundamental causes and interactions and served as the basis of the subsequent reliability assessment.

Nuclear Rocket Development Station (NRDS)

Major test facilities, including all the reactor test facilities, for the nuclear rocket program were located at the Nuclear Rocket Development Station (NRDS) on the Nevada Test Site (NTS). The NRDS was located on a desolate flat basin called Jackass Flats (Jackass Flats was named for some of the indigenous wildlife). Both reactors and complete engine assemblies were tested at the NRDS. Over time, three major test areas were built (see Fig. 24): Test Cells A and C and the Engine Test Stand Number One (ETS-I). These three test areas were connected by road and railroad to the R-MAD and E-MAD Buildings (Reactor and Engine Maintenance, Assembly, and Disassembly). Typically, a reactor or engine was assembled in one of the MAD buildings and then transported by the "Jackass & Western Railroad" to one of the test locations (see Fig. 24). This railroad was humorously referred to as the world's shortest and slowest. After a test, while the reactor or nuclear rocket engine was still highly radioactive, a heavily shielded railroad engine was used to tow it back to one of the maintenance, assembly and disassembly buildings for disassembly. The MAD buildings were immense "hot cells" where engineers, protected by thick concrete and steel shields, could perform robot-assisted disassembly of the radioactive reactors and engines.

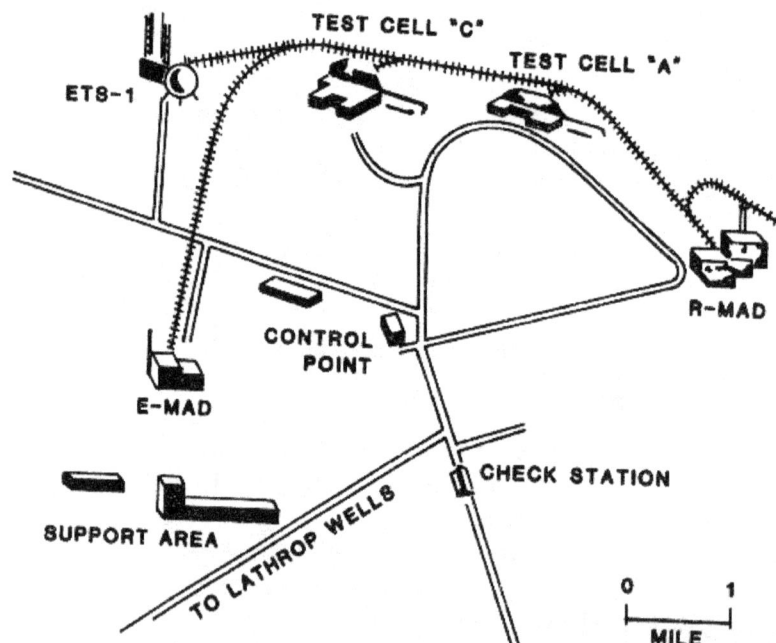

Fig. 24. Arrangement of facilities at the Nuclear Rocket Development Station. *Courtesy of DoE*

The reactor test facilities were designed to test a reactor in an upward-firing position (see Fig. 25); the engine test facility (see Fig. 26) tested engines (but not complete rockets) in a downward firing mode. During a nuclear rocket reactor test at the NRDS the desert basin became a literal inferno as the very hot hydrogen gas spontaneously ignited upon contact with the air and burned (with atmospheric oxygen) to form water. Unlike the familiar "pillar of fire" encountered with chemical rockets, during Earth based testing a nuclear rocket engine's exhaust is essentially invisible--except perhaps for any incandescent impurities and the thermal aberrations it produces in the surrounding air (see Fig. 27).

Fig. 25 The Pewee I reactor being transported to a test location at the NRDS by the "Jackass & Western Railroad." *Courtesy of NASA.*

Fig. 26. The ground experimental engine (XE) installed in Engine Test Stand No. 1 at the Nuclear Rocket Development Station in Nevada. *Courtesy of NASA.*

Fig. 27 The NRX-A6 engine being tested at the NRDS in December 1967. *Courtesy of NASA.*

Dual Mode System[16]

The thermal nuclear rocket can be modified to generate electric power along with propulsion. Such a configuration is called a dual-mode rocket system. One mode involves the normal propulsion function; the second mode involves the generation of electric power (at low to modest levels). Keeping the normal propulsion mode unchanged, the power production mode might be incorporated into a NERVA nuclear rocket engine design by isolating hydrogen flow through the structural support system from the reactor chamber and nozzle sections and continuously recirculating it (see Fig. 28). With this type of scheme, some 10 to 25 kilowatts of electric-power might be generated by a one MW-thermal rocket reactor (see Fig. 29). This particular dual-mode configuration used an organic Rankine cycle for electrical generation. In the 10 kW_e power production version, the heat rejection radiator could be located along the surface of the power module. For higher levels of electric power generation, additional radiator surface area would be required. The major rocket engine modifications needed to implement the power production mode in this fashion would be to add isolation valves for the low power gaseous hydrogen flow through the structural support system; to change the reactor dome and tie-tube core support plate and lines materials from aluminum to stainless steel in order to accommodate a wider range of operating temperatures; and to change the actuator materials from polyemid to ceramic materials. Engineering calculations indicate that over a two year lifetime there would be no significant degradation of the reactivity control margin.

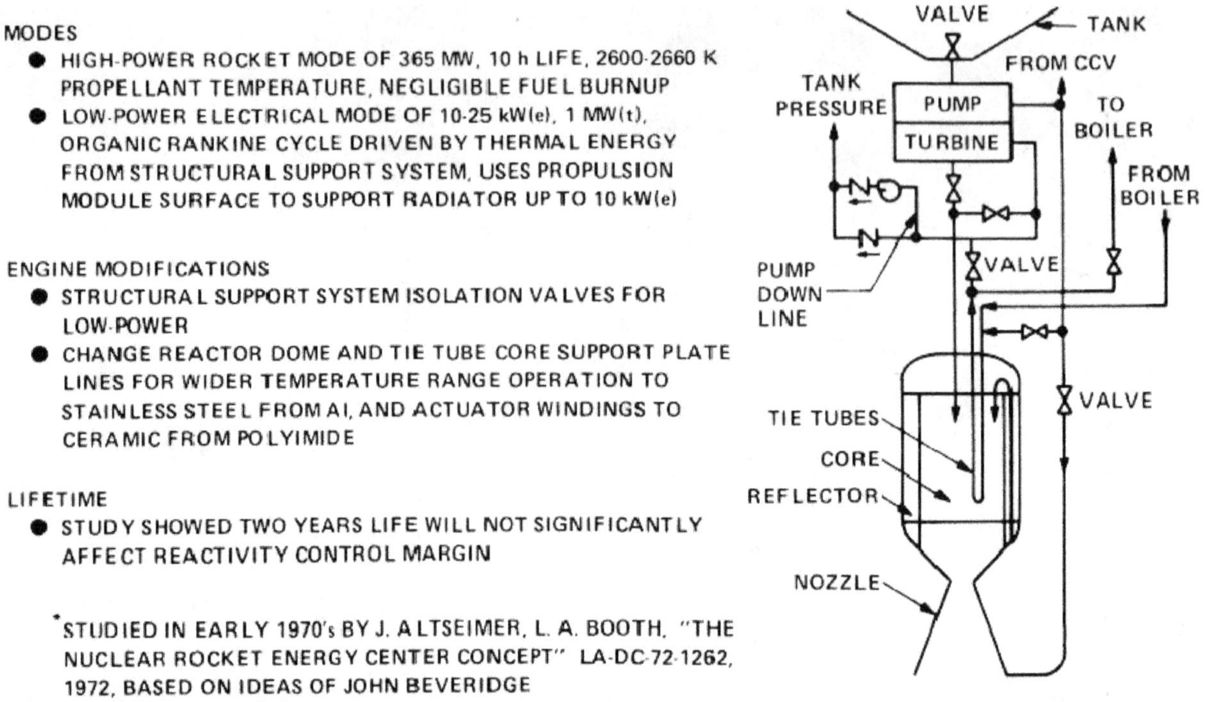
MODES
● HIGH-POWER ROCKET MODE OF 365 MW, 10 h LIFE, 2600-2660 K PROPELLANT TEMPERATURE, NEGLIGIBLE FUEL BURNUP
● LOW-POWER ELECTRICAL MODE OF 10-25 kW(e), 1 MW(t), ORGANIC RANKINE CYCLE DRIVEN BY THERMAL ENERGY FROM STRUCTURAL SUPPORT SYSTEM, USES PROPULSION MODULE SURFACE TO SUPPORT RADIATOR UP TO 10 kW(e)
ENGINE MODIFICATIONS
● STRUCTURAL SUPPORT SYSTEM ISOLATION VALVES FOR LOW-POWER
● CHANGE REACTOR DOME AND TIE TUBE CORE SUPPORT PLATE LINES FOR WIDER TEMPERATURE RANGE OPERATION TO STAINLESS STEEL FROM Al, AND ACTUATOR WINDINGS TO CERAMIC FROM POLYIMIDE
LIFETIME
● STUDY SHOWED TWO YEARS LIFE WILL NOT SIGNIFICANTLY AFFECT REACTIVITY CONTROL MARGIN
*STUDIED IN EARLY 1970's BY J. ALTSEIMER, L. A. BOOTH, "THE NUCLEAR ROCKET ENERGY CENTER CONCEPT" LA-DC-72-1262, 1972, BASED ON IDEAS OF JOHN BEVERIDGE
VALVE
TANK
FROM CCV
TANK PRESSURE
PUMP
TURBINE
TO BOILER
FROM BOILER
VALVE
PUMP DOWN LINE
VALVE
TIE TUBES
CORE
REFLECTOR
NOZZLE

Fig. 28. Dual-mode nuclear rocket. ***From L. A. Booth and J. H. Altseimer, LA-DC-72-1111.***

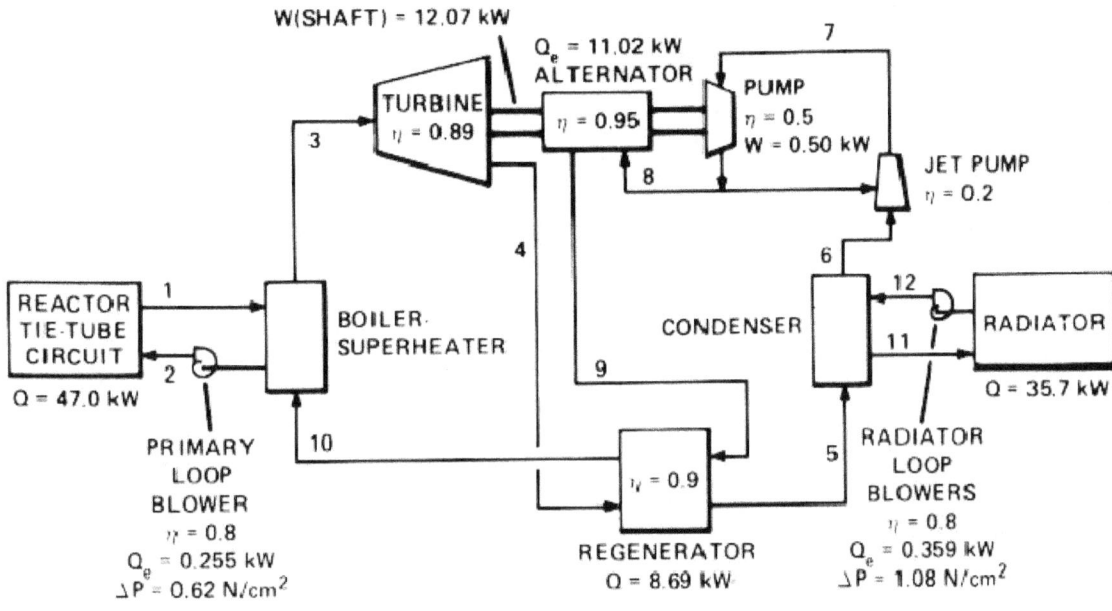

Fig. 29. Dual-mode nuclear rocket with a 10 kW$_e$ Rankine cycle. *From L. A. Booth and J. H. Altseimer, Report LA-DC-72-1111.*

Potential Rover/NERVA Growth

The Rover/NERVA programs demonstrated graphite/carbide based prismatic fueled nuclear thermal rockets. Advancements in fuels were continuing until the termination of the program; however, these advancements were not yet incorporated in the engine designs. Studies since the end of the program based on these advancements makes uses of these developments (see Table 7).

Table 7. Characteristics for 337 kN (75,000 lb$_f$) NERVA type engines.[17]

Parameters	1972 NERVA**	"State-of-the Art" NERVA Derivatives**				
Engine Flow Cycle	Hot bleed/Topping	Topping (expander)				
Fuel Form	Graphite	Graphite		Composite		Carbide
Chamber Temperature (K)	2,350-2,500	2,500	2,350-2,500	2,700		3,100
Chamber Pressure (psia)	450	500	1,000	500	1,000	1,000
Nozzle Expansion Ratio	100:1	200:1	500:1	200:1	500:1	500:1
Specific Impulse (s)	825-850/ 845-870	875	850-885	915	925	1,020
Engine Weight + (kg)	11,250	7,721	8,000	8,483	8,816	9,313
Engine Thrust/Weight (with internal shield)++	3.0	4.4	4.3	4.0	3.9	3.7

** Engine weights contain dual turbopump capability for redundancy.
+ Without external disk shield.
++ Thrust-to-weight ratios for NERVA/NDR systems are ~ 5-6 at the 250 klb$_f$ level.

Additional studies were performed on concepts called ENABLER. The ENABLER I concept is based on maintaining the NERVA fuel element dimensions. The fuel elements are fabricated using (U, Zr)C-Graphite composite material developed late in the Rover/NERVA program and tested in the Phoebus, Pewee and Nuclear Furnace reactor test programs. These exhibit improved corrosion resistance and allow higher operating temperatures. Zirconium-hydride moderator placed in the core support elements was demonstrated in the Pewee reactor. This increases the neutronic reactivity and thereby decreases the required uranium fuel loading. The ENABLER designs are designed to operate at a higher nozzle chamber pressure and temperature with a lower core pressure drop then the NERVA baseline engine. Fig. 30 is a schematic of the ENABLER design. [18]

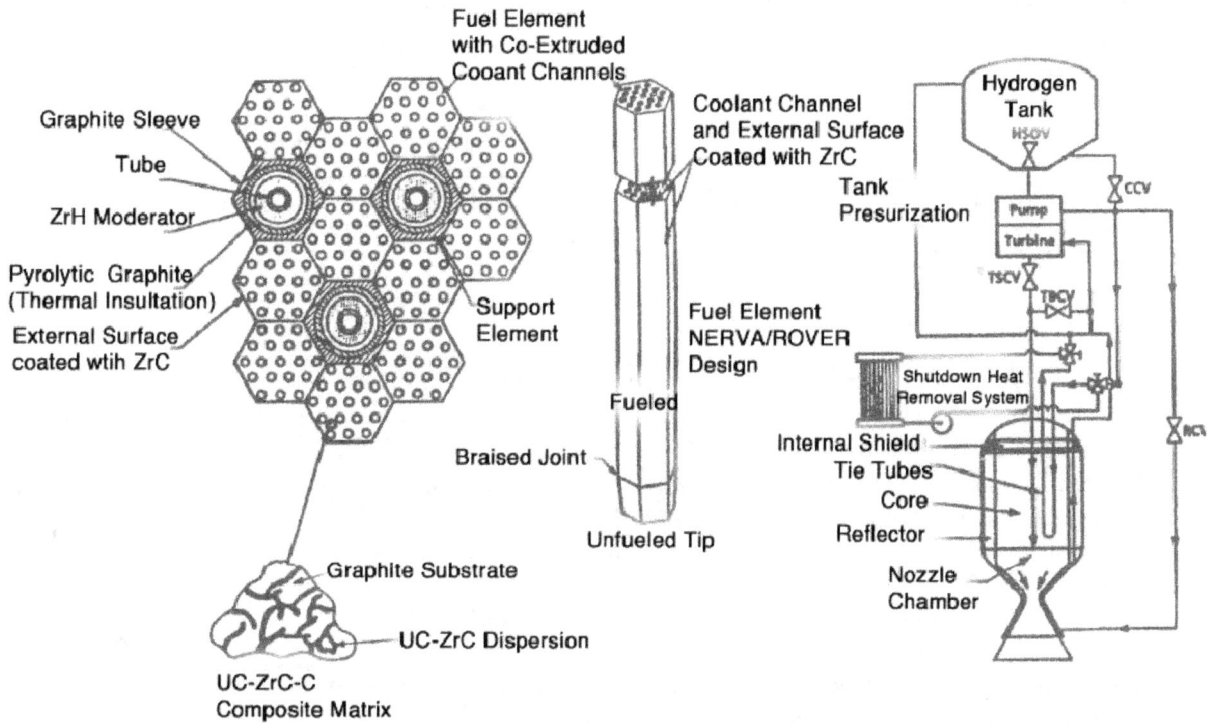

Fig. 30. Schematic of Enabler Engine and fuel structure. *From Lyman J. Petrosky, 1992.*

The engine uses a full flow cycle and other components under development in the NERVA program. ENABLER I can be sized for thrusts from 65 kN (15,000lb) to over 1,100 kN (250,000 lb). The chamber conditions are 2,700 K (4,860 R) and 7 MPa (1,000 psia). For a 335 kN (75,000 lb) thrust at a power level of 1,613 MW$_t$ the thrust to weight ratio is 59 N / kg (6.0 lb$_f$/ lb$_m$) without internal shield, and 50 N / kg (5.0 lb$_f$/ lb$_m$) with shield.[19]

ENABLER II is a further optimization of the design. The reactor design is the same as ENABLER I, but the fuel size is scaled to achieve optimal fuel performance. The scaling maintains similitude in the thermal and hydraulic profiles of fuel elements and of the core as a whole. This allows the power density for an element to increase while the thermal stress levels remain unchanged. Therefore, scaling maintains a strong linkage with the existing Rover/NERVA database while allowing departures from the previous fuel power density limitations. The higher power density translates into a smaller core and lighter engine at a given thrust level. The advantage of this scaling is seen in Table 8.[20]

Additional fuel development started in the NERVA/Rover program could further increase propellant temperature to 3,100 K and performance to a specific impulse of over 1,000 seconds (see Table 9).

Table 8. Thrust to Weight (F/W) comparison for ENABLER I and II.

Thrust Level		ENABLER I				ENABLER II			
		With shield		Without shield		With shield		Without shield	
kN	klb	N/kg	lb$_f$/lb$_m$	N/kg	lb$_f$/lb$_m$	N/kg	lb$_f$/lb$_m$	N/kg	lb$_f$/lb$_m$
130	30	34	3.5	41	4.2	48	4.9	64	6.5
220	50	40	4.1	48	4.9	58	5.9	75	7.7
330	75	50	5.0	59	6.0	69	7.0	90	9.2
440	100	53	5.4	63	6.4	74	7.5	96	9.8
560	125	55	5.6	66	6.7	80	8.2	105	10.7
670	150	57	5.8	68	6.9	83	8.5	109	11.1

Table 9. Improvements in specific impulse through advanced fuel developments.

Improvement	Specific Impulse (sec)	Risk
Binary ZrC-UC Carbide fuel (Chamber temperature = 3,100 K)	1,020	Medium
Graded fuel elements	1,030	Medium
Ternary UC-ZrC Carbide fuel (Chamber temperature = 3,300 K)	1,080	High

Summary

The nuclear Rover/NERVA rocket programs provides a very high confidence level that the technology for a flight nuclear rocket exist. These programs demonstrated power levels between 507 MW and as high as 4,100 MW$_t$. The Phoebus-2A test reactor achieved thrust levels of 930 kN (200,000 Ib$_f$) with a hydrogen flow rate of 120 kg / s (see Fig. 31). The Pewee test demonstrated the highest equivalent specific impulse, reaching 8,280 m / s (845 s) at an average coolant exit temperature of 2,550 K and a peak fuel temperature of 2,750 K. The Phoebus-2A system demonstrated the minimum reactor specific mass, a value of 2.3 kg / MW$_t$. The Pewee design had the highest core average power density at 2,340 W$_t$ / cm^3. NF-1 demonstrated a peak fuel power density of 4,500 W$_t$ / cm^3 and also achieved the longest accumulated time at full power for some 109 minutes. The XE' performed the greatest number of engine starts and restarts with 28. In summary, basic nuclear rocket research and technology was completed during the Rover/NERVA Programs. At the programs termination, the flight engine development program phase was just getting underway. To have developed a flight system, it would have been necessary to verify the flight reactor and engine design, perform duration testing, and conduct reproducibility testing. In summary, however, there appear to be no technical barriers to the development of a successful nuclear rocket.

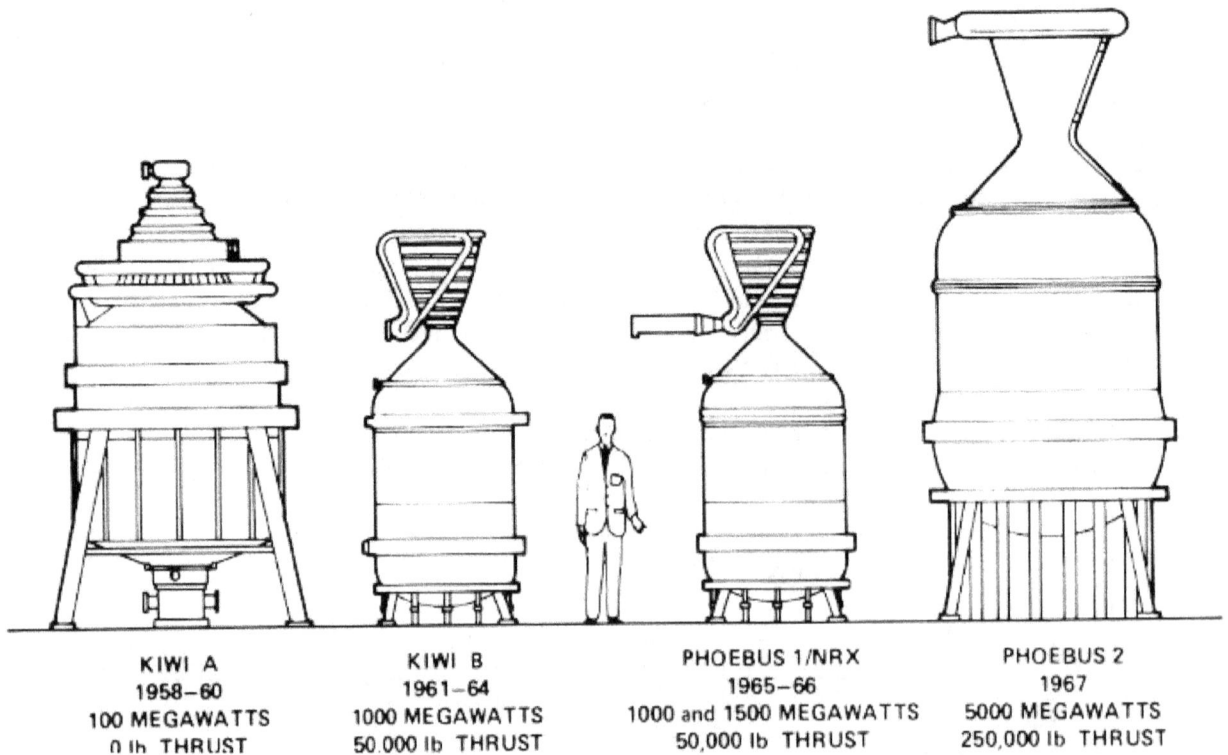

KIWI A	KIWI B	PHOEBUS 1/NRX	PHOEBUS 2
1958–60	1961–64	1965–66	1967
100 MEGAWATTS	1000 MEGAWATTS	1000 and 1500 MEGAWATTS	5000 MEGAWATTS
0 lb THRUST	50,000 lb THRUST	50,000 lb THRUST	250,000 lb THRUST

Fig. 31. Comparison of reactors tested in Rover Program. *From R. E. Schreiber, "The LASL Nuclear Rocket Propulsion Program," LASL Report LAMS-2036, April 1956.*

Bibliography

J. H. Altseimer, G. F. Mader, and J. J. Stewart "Operating Characteristics and Requirements for the NERVA Flight Engine," *J. Spacecraft, Vol. 8, No.7*, (July 1971).

J. H. Altseimer et al., "XE-Engine Systems Analysis Design Data Book," *Aerojet General Report No. RN-S-0289A*, (April 1967).

J. D. Balcomb, "Nuclear Rocket Reference Data Summary," *LA- 5057-MS*, (October 1972).

J. H. Beveridge, "Feasibility of Using a Nuclear Rocket Engine for Electrical Power Generation," *AIAA Paper 71-639, AIAA/SAE Joint 7th Propulsion Joint Specialist Conference*, Salt Lake City, Utah, (14-18 June 1971).

L. A. Booth and J. H. Altseimer, "Summary of Nuclear Engine Dual-Mode Electrical Power System Preliminary Study," *LASL Report LA- DC-72-111 I*.

D. Buden, "Operational Characteristics of Nuclear Rockets," *AIAA Paper 69-515, AIAA 5th Propulsion Joint Specialist Conference*, U.S. Air Force Academy, Colorado, (9-13 June 1969).

K. V. Davidson et al., "Development of Carbide-Carbon Composite Fuel Elements for Rover Reactors," *LASL Report LA-5005*, (October 1972).

J. DeStefano and R. J. Bahorich, "Rover Program Reactor Tests Performance Summary NRX-AI Through NRX-A6," *Westinghouse Astronuclear Laboratory Report WANL-TME-1788*, (July 1968).

R. Driesner, "Summary of Disassembly and Post-Mortem Visual Observations of the Kiwi-B4E-301 Reactor," *LASL Report LA- 3299-MS*, (July 1965).

F. P. Durham, "Nuclear Engine Definition Study Preliminary Report, Vol. I-Engine Description, Vol. II-Supporting Studies, Vol. III -Preliminary Program Plan," *LASL Report LA-5044-MS, Vol, I*, (September 1972).

W. J. Houghton and W. L. Kirk, "Phoebus II Reactor Analysis," *LASL Report LAMS-2840*, (April 1963).

M. T. Johnson, "NERVA Reactors," *Proc. ANS 1970 Topical Meeting*, Huntsville, Alabama, (28-30 June 1970).

L. D. P. King et .al., "Description of the Kiwi-TNT Excursion and Related Experiments," *LASL Report LA-3350-MS*, (August 1966).

W. L. Kirk, "Advanced Nuclear Rocket Technology," *Proceedings ANS 1970 Topical Meeting*, Huntsville, Alabama, (28-30 April 1970).

W. L. Kirk, "Nuclear Furnace-I Test Report," *LASL Report LA- 5189-MS*, (March 1973).

L. L. Lyon, "Performance of (U, Zr)C-Graphite (Composite) and of (U, Zr)C (Carbide) Fuel Elements in the Nuclear Furnace-I Test Reactor," *LASL Report LA-5398-MS*, (September 1973).

Misra, J. H. Altseimer, and G. D. Hart, "Hi-Flight Coolant Management Considerations for the NERVA Reactor Cooldown," *Nuclear Technology, Vol. 12*, (November 1971).

J. M. Napier, "NERVA Fuel Element Development Program Summary Report-July 1966 Through June 1972 Impregnation Studies," *Oak Ridge Y-12 Plant Report Y-1852, Part 2*, (September 1973).

M. Rice and W. H. Arnold, "Recent NERVA Technology Development," *J. Spacecraft, Vol. 6, No.5* (May 1969), p. 565.

J. C. Rowley, W. R. Prince, and R. G. Gido, "A Study of Power Density and Thermal Stress Limitations of Rover Reactor Fuel Elements," *LASL Report LA-3323-MS*, (July 1965).

R. E. Schreiber, 'The LASL Nuclear Rocket Propulsion Program," *LASL Report LAMS-2036*, (April 1956).

R. W. Schroeder, "NERVA-Entering a New Phase," *Astronautics and Aeronautics, (May 1968)*, p. 42.

J. M. Taub, "A Review of Fuel Element Development for Nuclear Rocket Engines," *LASL Report LA-5931*, (June 1975).

E. A. Warman, J. C. Courtney, and K. O. Koebberling, "Final Report of Shield System Trade Study," *Vol. I, Book I, Aerojet General Report S054-023*, S054-CP090290-F I.

Chapter 4

Russian Nuclear Rocket Development

Beginning in the late 1950s, the Soviet Union also performed innovative work on NTP. Experimental investigations and analytical designs were created including fuel elements with working fluid temperatures up to 3000 K, hydrogen turbo-pumping units, special high-temperature heat-insulating designs, moderators, beryllium reflectors, etc. Also, advanced schemes using nuclear fuel particles in a centrifugal forces in the vortex flow of gaseous working fluid and gaseous plasma reactors were examined. However, the primary efforts was on a direct thrusting rocket engine using H_2 propellant heated by solid carbide fuel elements. The goal was specific impulse in the 900-1,000 s level.[1] Following the Chernobyl accident,[2] the program was stopped in 1988.

Design and Experimental Test Results

The USSR Nuclear Rocket Engine (NRE) uses a heterogeneity principle of core modularity, unlike the homogeneous reactor designs of U. S. programs. In such a scheme a moderator material is placed separate from the fuel element (FE) containing the uranium. Fuel elements are thermally insulate from the moderator and placed in a housing. Combined they are a complete independent unit, named a fuel assembly (FA).

In the NRE design, fuel elements are made in the shape of a twisted ribbon (Fig. 1a) and assembled in bundles (Fig. 1b) with 6 to 8 integrated into each rod-type fuel assembly (Fig. 1c). These are inserted into channels in the propellant cooled zirconium hydride neutron moderator filling the reactor's core (Fig. 1d). [3] The fuel is high-temperature, corrosion-resistant ternary carbide (UC-ZrC-NbC and UC-ZrC-C) with a maximum operating temperature of about 3,200 K. The twisted ribbon surface-to-volume ratio is 2.6 times higher than that of the NERVA fuel elements. This enhances the heat transfer between the fuel and propellant. Power densities of up to 40 MW_t/ liter with minimum core mass characteristics of about 0.3 MW_t/ kg are claimed.[4]

One feature of the fuel assembly design is the ability to change the axial physical profile. The UC-ZrC-NbC and UC-ZrC-C fuel elements are placed up-stream while the UC-ZrC-NbC are downstream. Because the high temperatures are localized to the fuel element bundles, the rest of the core operates at much lower temperatures (< 1,073 K). This permitted the use of common materials in the core's structure. In particular, low melting point neutron moderators, such as zirconium and lithium hydrides, are possible. The fuel assemblies can be changed in both the radial and circumferential directions for radial profiling. Also, the propellant mass flow rate through each fuel assembly can be controlled, making hydraulic profiling possible.

This configuration allows one to assess the NRE's performance as a single rod fuel assembly. "Since there is no need to test the whole reactor core to assess the NRE's performance, such a reactor has never been built so the performance results quoted are, indeed, the results for a single rod assembly tested in a research reactor core."[5] In about 15 tests in a research reactor EWG-1, measurements gave a maximum hydrogen temperature of 3,100 K. These tests correspond to a specific impulse of 925 s, a total lifetime with 10 consecutive restarts of 4,000 s, and a thermal power density of 10 MW_t/ L. Tests also demonstrated 4,000 hr operation at 2,000 K.

Multiple test reactor facilities were built with which to test individual components under development, like fuel elements and moderator modules. The IVG-I experimental reactor for fuel element testing is shown in Fig. 2.[6]

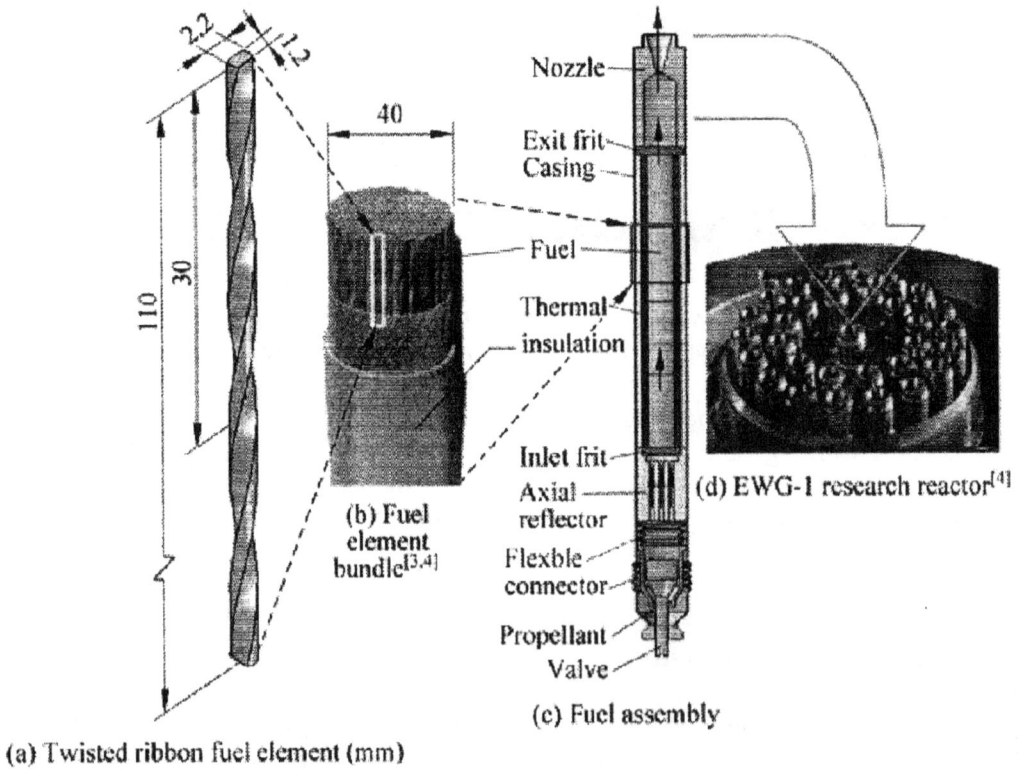

(a) Twisted ribbon fuel element (mm)

(b) Fuel element bundle[3,4]

(c) Fuel assembly

(d) EWG-1 research reactor[4]

Fig. 1. Russian NRE concept.

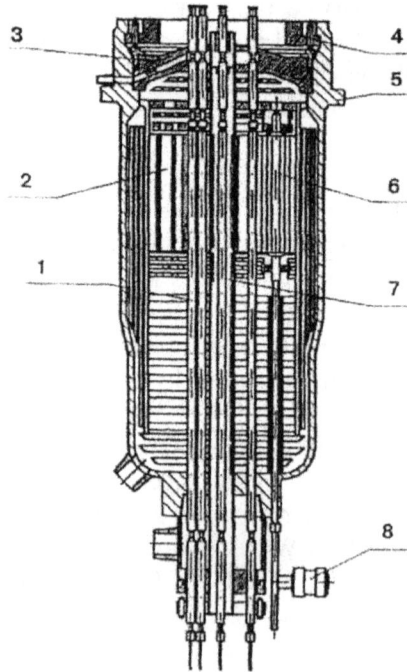

Fig. 2. Russian IVG-1 schematic. (1) Fuel Assembly; (2) reflector; (3) lid; (4) lock; (5) vessel; (6) control element; (7) loop channel; and (8) drive.

In the 1970 to 1988 time period, some 30 firings were conducted in research reactors. These tests explored thermal power levels up to 230 MW$_t$ and propellant mass flow rates up to 16.5 kg / s. The maximum power density in the fuel composition reached 25 MW$_t$ / L, the uranium enrichment being 90% with the U^{235} load varied from 6.7 to 15.9 kg. The radioactive product released from the reactor core through the exhaust was measured to $\leq$ 1% (by mass).

In the 1975 - 1989 time period, the reliability of the carbide fuel elements was proven by successful testing 330 fuel element bundles. In these tests, maximum power densities reached 35 MW$_t$ / L. At 3,100 K, the temperature transient rate reached 1,000 K / s with up to 12 thermal cycles tested.

The data from these tests form the bases for the RD-0410 ~ 35 kN thrust design. A full scale prototype of RD-0410 (see Fig. 3a and 3b) was built and tested using electric heaters. The prototype has the performance characteristics listed in Table 1. The RD-0411 data is from A. Borisov[7]. Hexane has been suggested as a propellant additive to reduce the fuel element erosion by the hot hydrogen. One important safety feature of the design is that the complete NRE assembly can be tested on the final phase of preparation before flight by replacing fissile material with simulators.[8]

Fig. 3a. RD-0410 NTP Engine. *Credit: Dietrich Haeseler*

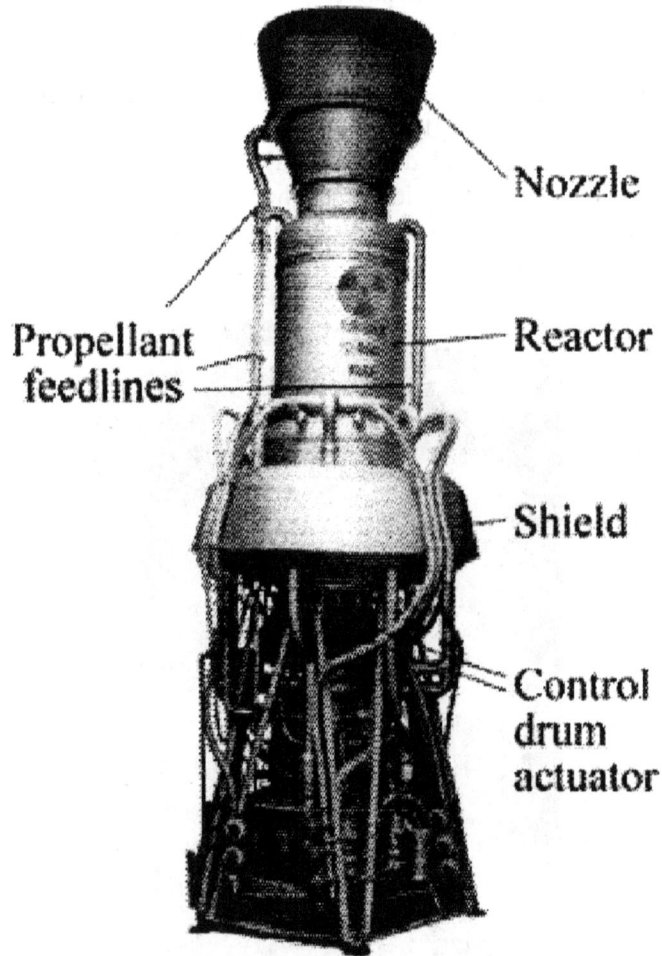

Fig. 3b. RD-0410 with call outs.

Table 1. Performance of Russian NREs.

Feature	RD-0140	NPPS
Thrust (vacuum) (kN)	35.28	68
Propellant	H_2 + Hexane	H_2
Propellant flow rate (kg/s)	~4	~7.1
Specific impulse (vacuum) (s)	~900	920
Core outlet temperature (K)	3000	2800-2900
Chamber pressure (10^5 Pa)	70	60
Fuel U^{235} enrichment (%)	90	90
Fuel composition	(U,Nb,Zr)C	U-Zr-C-N
Fuel element form	Twisted ribbon	Twisted ribbon
Generated electric power (kW)	N/A	50
Working fluid for power loop (%, by mass)	N/A	93% Xe +7% He
Maximum temperature for power loop (K)	N/A	1500
Maximum pressure for power loop (10^5 Pa)	N/A	9
Working fluid flow rate (kg/s)	N/A	1.2
Thermal power (MW)		
Propulsion mode	196	340
Power mode	N/A	0.098
Core dimensions (mm)		
Length	800	700
Diameter	500	515
Engine dimensions (mm)		
Length	3700	No data
Diameter	1200	available
Lifetime		
Propulsion mode (h)	1	5
Power mode (year)	N/A	2
Mass (kg)	2000[*]	1800[**]

Note: N/A, not applicable; [*], including radiation shield and adapter;
[**], reactor mass.

The engine cycle considered several turbopump configurations. The selected cycle is a closed cycle configuration with turbopump (TPU) and a gas generating heat release assembly (HRA) This permits independent selection of turbine components and allows as high a turbine temperature as desired. U^{235} is added to the moderator in order to increase the gas generator energy. Fig.4 is a schematic of the selected NRE engine. Reactivity control is by means of twelve drums along the reactor periphery.

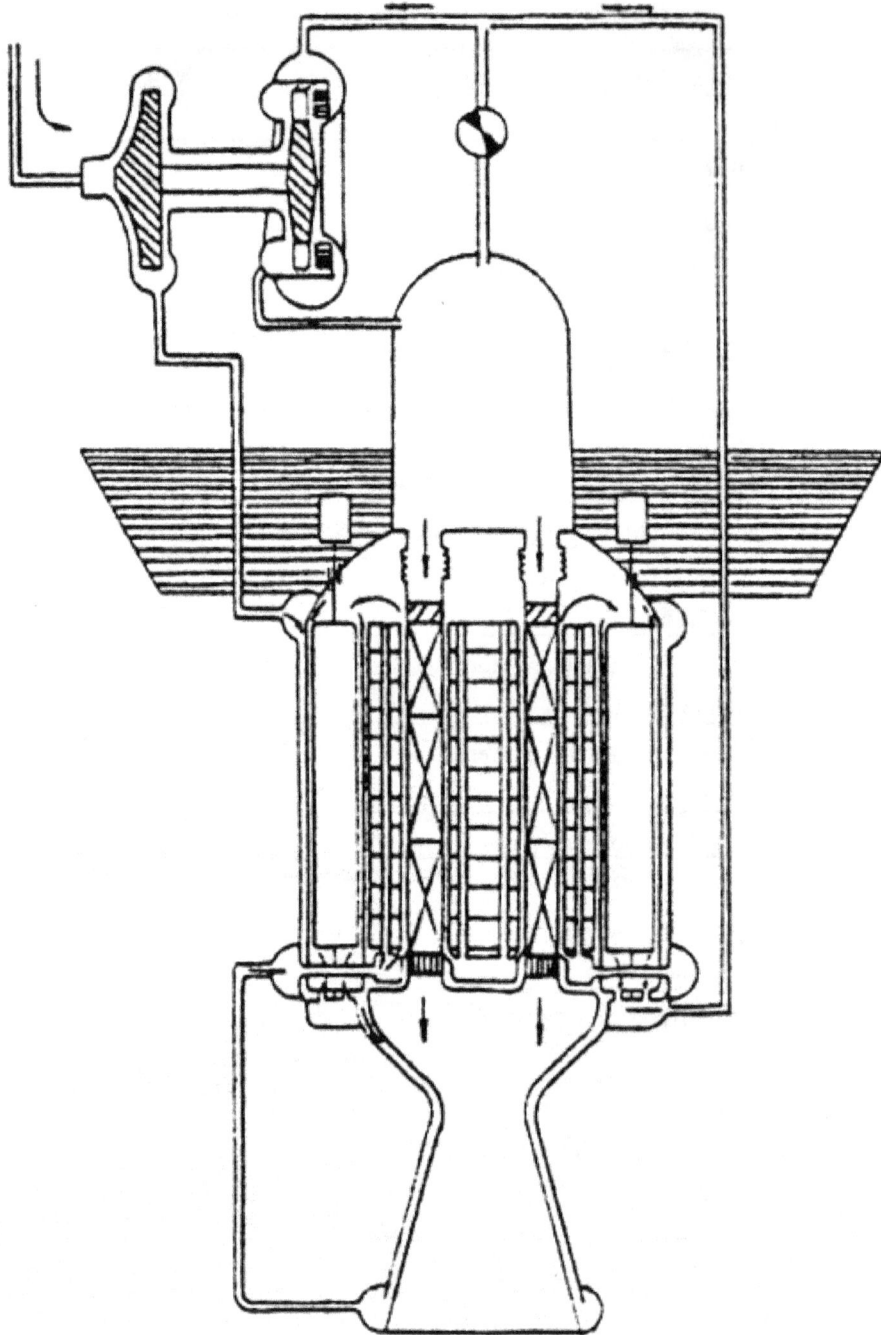

Fig. 4. NRE schematic.

Maximizing the mass-averaged working fluid temperature at the reactor exit is achieved by lowering the uranium concentration toward the reactor exit. The required thermal neutrons distribution along the reactor active zone is formed by placing in the "cold" reactor end the beryllium reflector moderator. The moderator thickness value can be considered infinite from the point of view of neutron physics. On the "hot" reactor end, the conditions are created for enhanced thermal neutron leakage. Fig. 5 presents typical distributions of average working fluid temperature and average fuel element wall temperature along the reactor active zone.

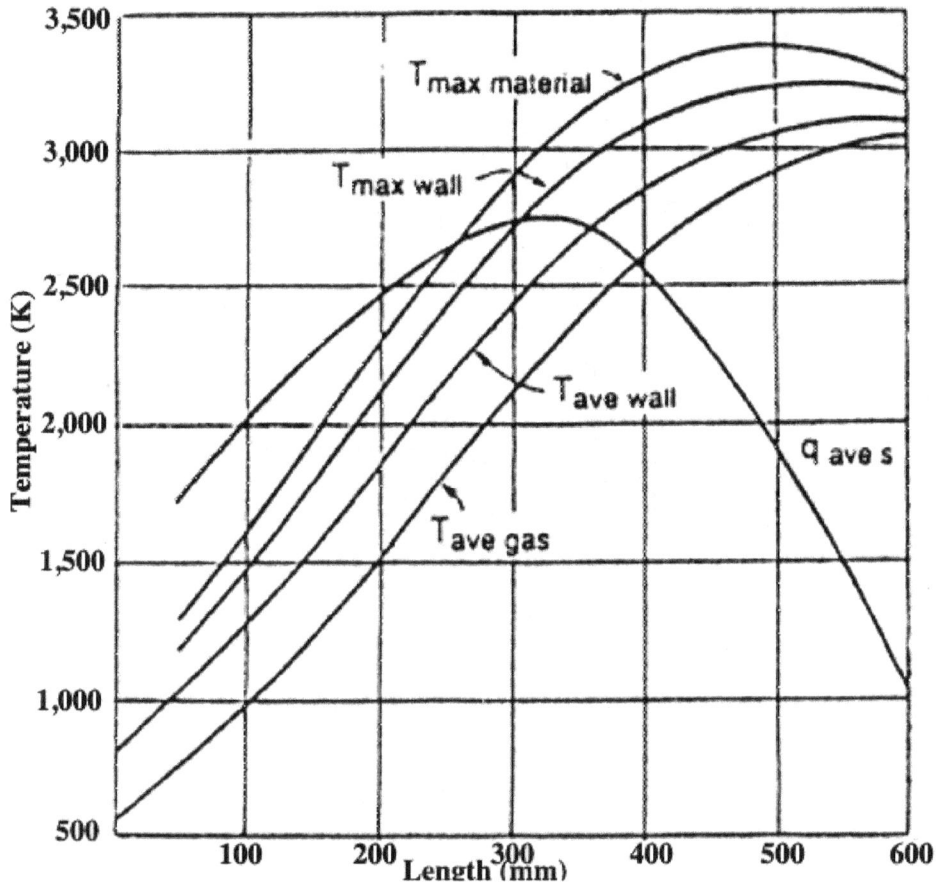

Fig. 5. Average working fluid temperature and average wall temperature variations along an active zone length.

The standard reactor has the heating sections with fuel elements, thermal insulation around the heating sections, thermal insulation to minimize heat losses from the reactor core, a support structure which resists the force due to hydraulic pressure, and reflector assembly with control drums. This minimize fast neutron leakage from the reactor to the payload and minimize the reactor shielding mass to protect components in the payload from radiation damage. A ratio of active zone length to diameter is chosen as two.

Results of studies to modify the NPR concept to include a Brayton cycle power loop are also shown in Table 1 in the column entitled NPPS. These parameters are from N. N. Ponomarev-Stepnoy, et al[9] Fig. 6 is a schematic of this concept. A mixture of xenon and helium is the suggested working fluid for the power mode. The chemical stability of recently developed carbide-nitride nuclear fuel has been confirmed during 100 hour tests at ~ 2,800 K, so this fuel is used in place of the initial carbide design. The addition of the Brayton cycle power loop provides a continuous source of reactor electrical energy. Operating the reactor continuously reduces the thermal stresses in the reactor since the engine is "pre-heated", minimizes large thermal cycles since there is no prolonged, deep "cold soak" of the engine, allows rapid reactor restart in case of an emergency, minimizes the "decay heat removal" propellant penalty by rejecting low power, decay heat through the power system's space radiator, and supplies hot, gaseous hydrogen for propellant tank pressurization and possible high specific impulse attitude control and orbital maneuvering systems.

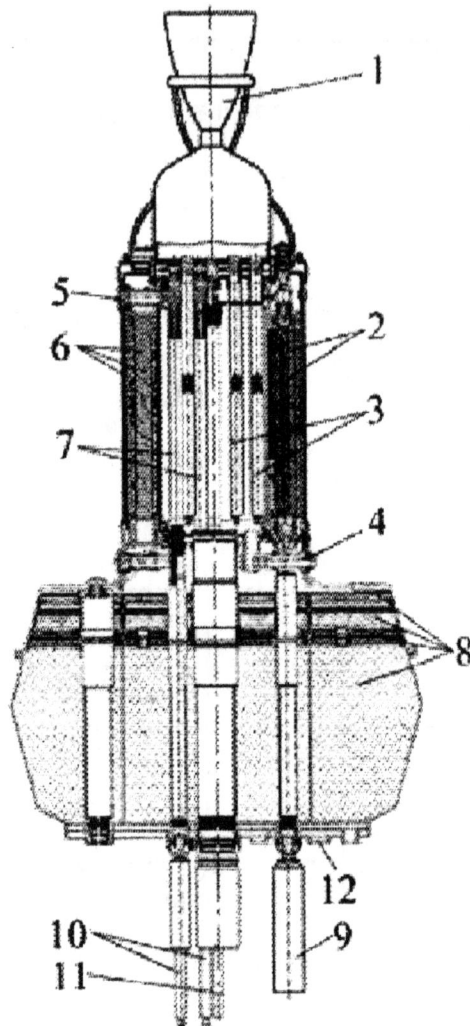

Fig. 6. NPPS schematic. (1) Nozzle; (2) Rotary drum; (3) Fuel assembly; (4) Outlet to turbo-generator; (5) HE+Xe mixture supply from turbo-generator; (6) Radial reflector; (7) Preheating fuel assembly; (8) Radiation shield; (9) Rotary drum actuator; (10) Radiation safety bar actuator; (11) Nozzle plug actuator; (12) H$_2$ inlet.

Summary

The Russian nuclear thermal rocket development has concentrated on some very unique design features. The reactor is a heterogeneous type with the zirconium hydrate solid moderator and beryllium reflector. The fuel elements are made in the shape of a twisted ribbon and assembled in bundles that are inserted into channels in the propellant cooled zirconium hydride neutron moderator. Control is by means of twelve drums in the side reflector. The turbopump unit is closed-cycle arrangement.

A method of autonomous testing of fuel elements was developed Experimental data indicates that hydrogen temperatures of 3,100 K, corresponding to a specific impulse of 925 s, has been demonstrated. A prototype NRE RD-140 was built and electrical heated. However, no full scale engine test has been run with a nuclear fueled reactor to demonstrate integrated engine system performance. The ribbon fuel appears to be difficult to fabricate, susceptible to brittle failure, and prone to high fission gas release.[10]

Chapter 5

Particle-Bed Reactor Nuclear Rockets

Commercial High Temperature Gas-Cooled Reactors (HTGR) developed a particle size fuel that form the bases for the particle-bed reactor nuclear rocket. A typical fuel is made up of particles 700 μm in diameter, each with a 300-μm kernel of zirconium carbide that contains several percent of uranium carbide. Superior heat transfer of the small particles allows potentially higher reactor outlet temperatures of 3,000 K and improved specific impulse approaching 1,000 seconds. In 1986, particle-bed concepts were being studied for use as an orbital transfer rockets.[1]

In 1987, the Strategic Defense Initiative established a program entitle Timberwind to develop a Particle-Bed Rocket (PBR) engine for propelling long-range anti-missile interceptors. Small PBR-powered anti-missile interceptors with burn-out velocities in excess of 7 km / s are capable of placing kinetic kill vehicles on ICBM-like trajectories. This would permit intercepts at ranges in excess of 5,000 km, approximately 18 minutes after launch. Design and experimental work continued until 1991. The Timberwind concept promised significant reductions in system mass over solid-core reactors.

The goals of the Timberwind program focused on the development of a PBR engine with the following characteristics:

Power Density: 40 MW$_t$ / liter
Thrust/Weight Ratio: 30:1
Exhaust Temperature: 2,750 K
Specific Impulse: 869 seconds
Thrust: 178 kN
Run Time: 100 seconds
Restart Capability: No
Throttlable: No

Subsequently, the Air Force evaluated the PBR concept as an upper stage for military and other space launch vehicles. The goal here was to dramatically reduce cost, increase reliability and operability associated with routine access to space. However, the program was terminated in 1992.

Particle-Bed Reactor Description

High performance particle fuel is required to realize the potential performance of a PBR nuclear thermal propulsion system. The selected fuel particle for optimum performance is a coated solid carbide particle. This selection is based on reaching the high temperatures for 1,000 second specific impulse performance and to achieve maximum fission product retention capability. The design comprises a mixed uranium-refractory metal carbide fuel kernel coated with a refractory metal carbide.

In the PBR design the fuel particles are held between two porous cylindrical frits to form an annular fuel element. The basic building block of the PBR design is a fuel element with packed fuel particles housed in a frit. The fuel particle bed contains millions of these tiny fuel particles.[2] Fig. 1 is a schematic of the fuel element arrangement with a moderator block. The coolant flows axially through the moderator, then enters the plenum surrounding the cold frit. The flow through the cold frit, fuel bed and hot frit is radial. The hydrogen temperature when it leaves the moderator is about 150 K Finally, the hot coolant leaves the fuel element by flowing axially out through the element. The hot gas flows along a central channel of each element to a common outlet plenum and nozzle. A 200 MW$_t$ design, shown in Fig. 2, has 19 such elements.[3]

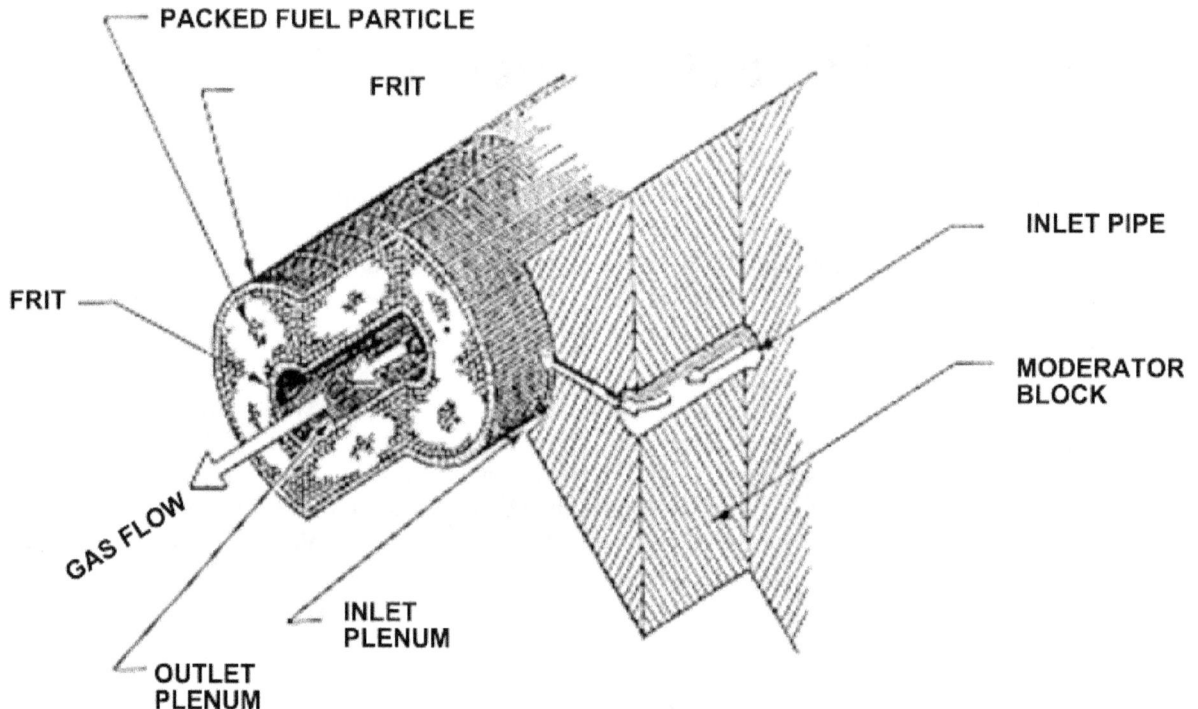

Fig. 1. Baseline fuel element and moderator block. *From F. L. Horn, J. R. Powell, and O. W. Lazareth, 1987.*

The large heat transfer area (~ 100 cm^2 / cm^3) in the elements allows very high bed power densities (10 MW$_t$/ L) and compact, lightweight reactors. The fuel particles have a diameter of ~ 500 μm. These minimize the local temperature difference between the particles and coolant to about 50 K and permit fast startup capability. Full power can be reached in several seconds.

An advantage of the PBR design is that only a small portion of the core is at high temperatures. The pressure vessel, control drums, core moderator, reflector, and cold frit are at low temperatures of 300 K or less. This simplifies the design and increases reliability. The high temperature portions of the reactor include the hot frit in each element, the part of the fuel bed towards the propellant exit, the outlet plenum and nozzle. In the fuel bed, radial temperatures rise in an approximately linear manner. Thus, only about a third of the bed is above 2,000 K.

The PBR reactor shown in Fig. 2 is a thermal reactor, fueled with enriched ^{235}U (93.5 %) and moderated by beryllium (Be). The Be moderator contains channels for 19 fuel elements, each 6.4 cm in diameter. Six additional channels contain launch poison safety chains. These remain in place until orbit is achieved, and are withdrawn just before reactor startup. Coolant enters the core through the launch poison guide tubes in the Be moderator. It then flows through distribution holes in the Be to a thin annular plenum around each fuel element. Next, it flows radially inwards through the outer cool frit, the packed fuel particle bed, and the hot inner frit (see Fig. 3). The cooled portions of the core that include the moderator, structure, and cool frit are maintained at low temperature by the incoming propellant.

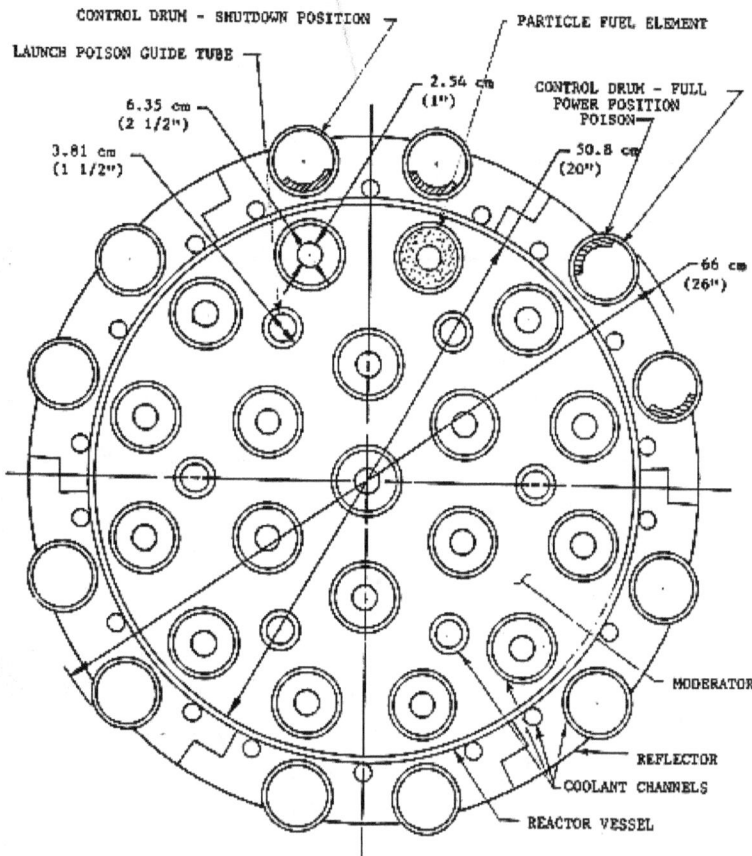

Fig. 2. Orbit transfer vehicle cross section. *From F. L. Horn, J. R. Powell, and O. W. Lazareth, 1987.*

As seen in Fig. 4, the heated propellant then exits along the central channels of each element to the common outlet plenum and nozzle. An aluminum pressure vessel is located between the core and the reflector. The rotating control drums our outside the pressure vessel and cooled by cold propellant before it enters the pressure vessel.

Table 1 summarizes key parameters for the reactor design. The reactor is designed to use either hydrogen or ammonia as the coolant. Total power output is 200 MW$_t$, producing a thrust of 50,000 N with hydrogen coolant and 80,000 N with ammonia coolant. The reactor assembly mass is 600 kg.

Table 1. Key reactor parameters for PBR reactor.

Core Diameter/Length	50/50 cm
Number of Fuel Elements	19
Fuel Element OD/ID/Length	6.5/1.9/71 cm
Hot Frit Composition/Thickness	ZrC Coated Carbon-Carbon/3.0 mm
Cold Frit Composition/Thickness	Stainless Steel/1.5 mm
Fuel Particle OD/Coating Composition	500 μm / ZrC
Fuel Particle Kernel OD/Composition	200 μm / UC-ZrC
Fuel Particle Buffer Layer Thickness/Composition	75 μm / PyC (50% D.F.)
Fuel Particle Sealer Layer Thickness/Composition	35 μm / PyC
Be Moderator OD/Length	50.8/66 cm
Fuel Bed Volume	15 L
Axial Element Reflector Composition/Thickness	Graphite/7.6 cm
Reactor Vessel Composition/Thickness	Aluminum/1.6 cm

Number/Composition/ID of Launch Poison Tubes	6/Aluminum/2.3 cm
Launch Poison Composition	B$_4$C Beads on string
Radial Reflector Composition/Thickness	Beryllium/5.2 cm
Number/Diameter of Control Drums	12/6.0 cm
Control Drum Composition	Beryllium/ B$_4$C/(120^0 segment)
Fuel Element Mass (19)	125 kg
Beryllium Moderator Mass (includes Axial Reflector)	145 kg
Beryllium Radial Reflector Mass	70 kg
Reactor Vessel Mass	110 kg
Control Drum Mass (including Drive)	95 kg
Launch Poison Mass	15 kg
Miscellaneous	40 kg
Total Reactor Assembly Mass	600 kg

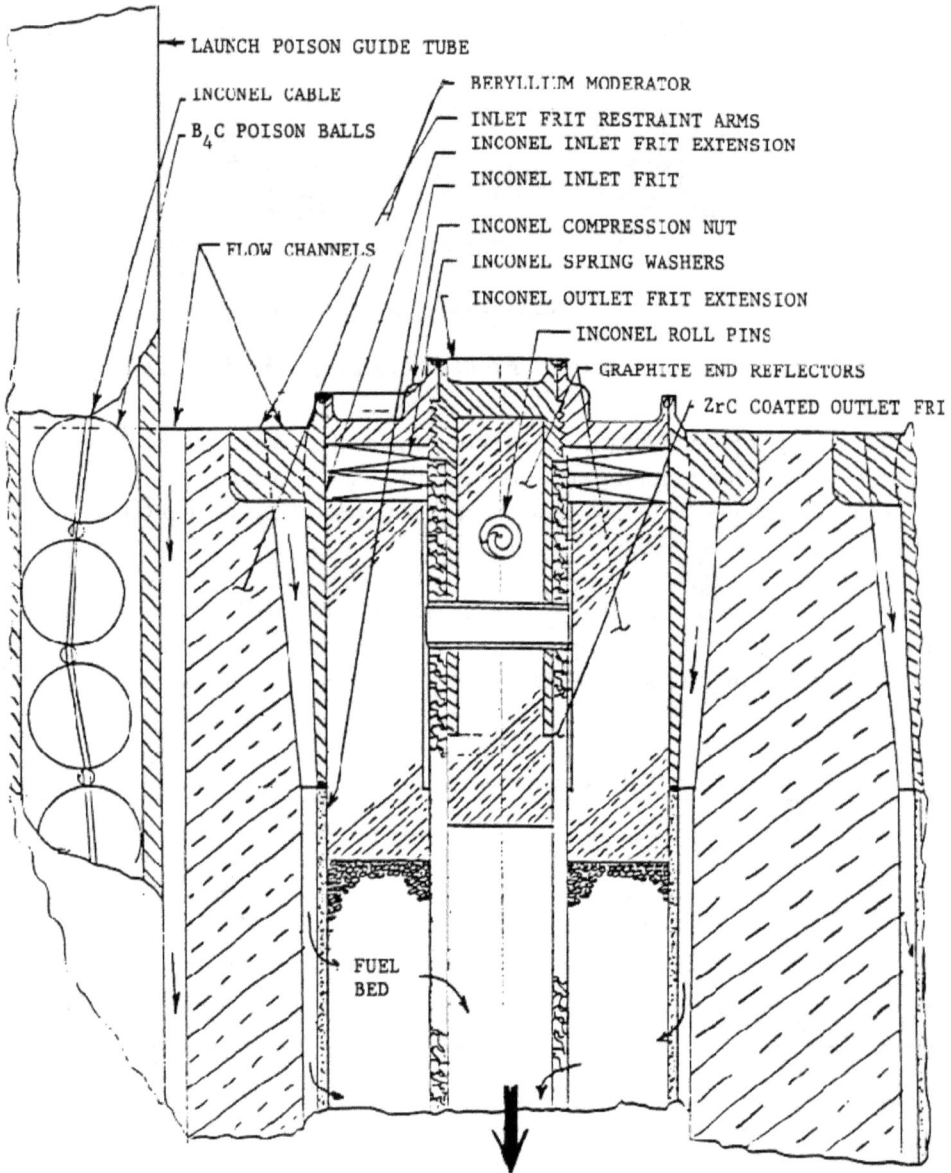

Fig. 3. Cross section of fuel element and moderator inlet. *From F. L. Horn, J. R. Powell, and O. W. Lazareth, 1987.*

BOOK 2
NUCLEAR THERMAL PROPULSION SYSTEMS

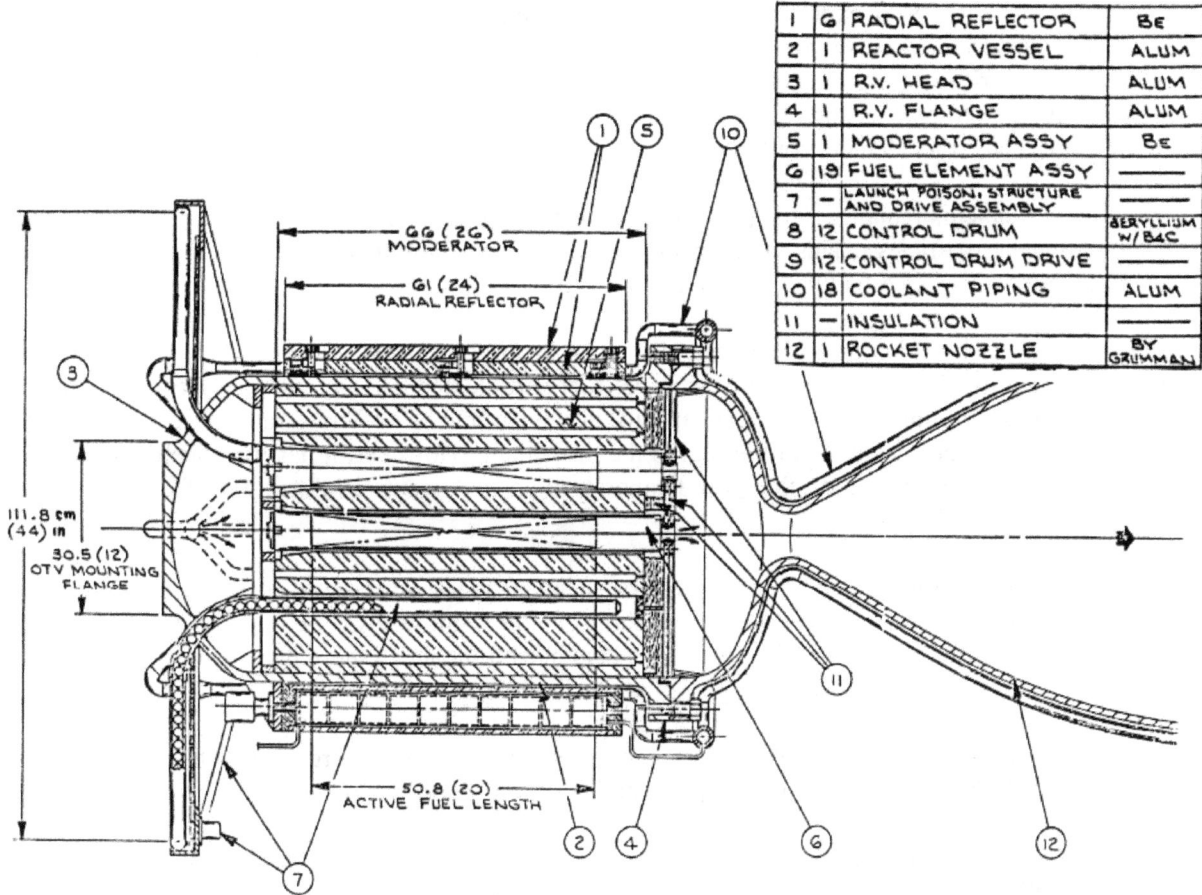

1	G	RADIAL REFLECTOR	BE
2	1	REACTOR VESSEL	ALUM
3	1	R.V. HEAD	ALUM
4	1	R.V. FLANGE	ALUM
5	1	MODERATOR ASSY	BE
G	19	FUEL ELEMENT ASSY	———
7	—	LAUNCH POISON, STRUCTURE AND DRIVE ASSEMBLY	———
8	12	CONTROL DRUM	BERYLLIUM W/ B4C
9	12	CONTROL DRUM DRIVE	———
10	18	COOLANT PIPING	ALUM
11	—	INSULATION	———
12	1	ROCKET NOZZLE	BY GRUMMAN

Fig. 4. Orbital transfer vehicle particle bed reactor assembly elevation view. *From F. L. Horn, J. R. Powell, and O. W. Lazareth, 1987.*

Technology Status[4]

Significant tests were performed on particle bed reactor designs during the Timberwind program.

Critical experiments were conducted at the Sandia Pulsed Reactor to support physics experiments with a zero-power PBR reactor core. These were used to benchmark neutronics simulations codes.

A baseline particle fuel design was based on technology from the HTGR program. Fuel particles were tested in a series of experiments called Pulsed Irradiation of PBR Fuel Element or PIPE. The fuel tested in the PIPE program consists of three non-fuel layers surrounding a fuel-material kernel. The kernel was composed of UC_x and fabricated by an internal-gelation technique. The value of x can be varied with a value of 1.65 used.[5]

The PIPE fuel particle tests were conducted at 5% of the scale of an operational system in the Sandia National Laboratory Annular Core Research Reactor (ACRR). These tests indicated that the PBR fuel element is a highly complex thermo-mechanical system, which is vulnerable to disruption at low power densities and low hydrogen flow rates. The tests also demonstrated an increase in propellant flow resistance during power operations, and a need for further understanding of local flow anomalies.

PIPE-1, conducted in October 1988, consisted of five cycles at increasing power levels, with a 1,600 K average and 1,900 K peak outlet temperatures. Power density levels reached 1.5 to 2.0 MW_t / liter. No off-normal results were observed

PIPE-2, conducted in July 1989, experienced significant anomalies. The experiment was terminated after 24 seconds of operation as a result of over-temperatures of the fuel element. The cause was believed to be blockages in the cold frit by foreign material in the motor. Extensive fuel particle fractures resulted, leading to frit clogging by particle fragments and release of fission product inventories. This led to a major redesign effort to develop a more robust fuel element configuration.

Major unresolved problems remain. One problem is debris clogging of the very small (micron size) holes in the frit supporting the particles. This resulted in localized redistribution of propellant flow, which led to thermal anomalies that damaged fuel particles and other structural elements.

Controlling the flow distribution throughout the reactor is another problem. Anomalous propellant flow and thermal profile conditions leads to local hotspots in the fuel particle bed and can ultimately cause fuel particle and structural damage or failure. The low heat capacity of the particle bed offers an attractive characteristic of more efficient thermal heat transfer from the fuel to the propellant. However, it has the potential of increased risk of melting of fuel particles and other low temperature core elements in the event of over-temperature events.

The low heat capacity of the PBR core requires very stringent reactivity control to avoid power overshoot during reactor startup. Failure to prevent power overshoots could result in structure material damage from high temperatures and fuel particle melting. Reactor controllers need to be much more responsive to control a PBR reactor.

Tests data to date have demonstrated fuel element power densities that are low relative to those of operational systems. Extensive testing is needed at more representative power densities.

As a result of the PIPE testing, an advanced fuel particle development was undertaken to achieve a mixed mean outlet propellant temperature of 3,000 K. The criteria are that the particle should remain solid during operation and that the fuel particle design should eliminate free carbon from the particle. The criteria for the fuel particle not to become even partially molten during operation is based on the performance during the PIPE experiments. The criteria to eliminate free carbon is a result of free carbon causing eutectic melting problems and may be a lost to the propellant to a significant degree during operation at the temperatures required to achieve a mixed mean outlet temperature of 3,000 K. To meet these criteria, a uranium-containing refractory metal carbide kernel is used bare or coated with one or more layers of refractory-carbide material. The fuel concept that would seem to meet the criteria for a PBR fuel is a coated mixed-carbide advanced fuel particle. Near-term, using a (U,Zr)C kernel coupled with a NbC (or ZrC) coating (Fig. 5), it will be possible to fabricate fuel with a high solidus temperature and fission-product-retention capability. Longer term, coated (U,Zr,Nb)C particle fuel will produce a fuel with the maximum operation temperature attainable.[6]

Several analytical models have been developed to understand flow stability.[7, 8] In a system with multiple heated flow channels cooled by a flowing gas, there is the possibility for temperature and flow instabilities. The instability mechanism can be described as: (1) parallel channels connected at plenum regions are stabilized at some inlet temperature and pressure and outlet pressure; (2) a perturbation in one channel causes the temperature to rise, thereby increasing gas viscosity and reducing gas density; (3) because the pressure drop is fixed by the plenum regions, the mass flow rate in this channel would then decrease; (4) the decrease in flow rate reduces the ability to remove the energy added, thus the temperature increases; (5) this process could continue until failure of the element. These sorts of instabilities are usually associated with low flows and heat rates above limiting values, mainly during rocket engine startup and shutdown. Analysis indicates that the cold frit flow resistance needs to be large enough to create an element pressure drop in which the fuel region pressure drop is small by comparison. Stability mapping generated by using the fuel region pressure drop as the criterion indicates that higher flows are necessary in order to maintain stability. Also, the element axial heat deposition shape can have a negative impact on stability.[9] Analytical studies can help in the design process; however, a full experimental engine test will be needed to verity flow stability.

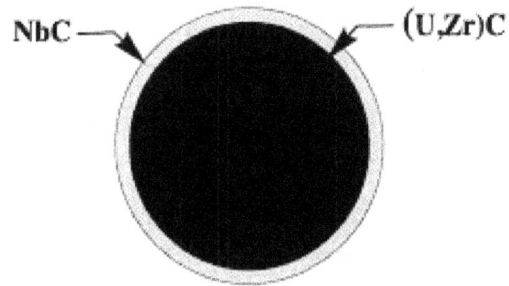

Fig. 5. Near-term fuel particle design. *From DeWayne L. Husser, et al., 1994.*

Summary

The pebble-bed nuclear thermal rocket potentially has some very significant advantages over other solid fueled reactor systems. This includes specific impulses approaching 1,000 seconds compared to NERVA's 875 seconds for the small engine design and significantly lower thermal rocket mass of about one-fifth of NERVA technology. However, the technology is still in the experimental stage with significant advances needed in fuel particle development, fuel elements including frits design, flow stability, fuel compaction and clogging of fuel particles and system controls. The 1987 - 1991 experimental program helped define the state of the technology and provides clues to solutions to these problem areas.

Chapter 6

Additional Nuclear Propulsion Concepts

Along with the nuclear propulsion concepts already discussed, a number of alternate nuclear configurations have been studied. Some of these use other solid fuel element forms such as pellets and wire wrap and have the same material restrictions as the previous concepts. Others, such as a molten core reactors could operate at temperatures of 5,000 K and specific impulse of 25,500 m / s. Experiments have been performed to support gaseous core reactors that operate to 20,000 K and specific impulse of 66,000 m/s. Concepts have also been proposed for direct fission/fusion, thermonuclear fusion and annihilation of matter with specific impulses greater than 1.3×10^7 for possible missions outside our solar system.

Pellet Bed Reactor Concept[1, 2]

The Pellet Bed Reactor (PeBR) is a fast-flux (peak neutron energy = 1 - 50 keV), hydrogen cooled reactor fueled with graphite-ZrC coated, spherical fuel pellets, 1.0 cm in diameter. The fuel pellets in the core are self supported, eliminating the need for internal core structure. This both simplifies the design and reduces the size and weight of the core. With the absence of internal core structure, reactor fueling can be performed at the launch site or even in orbit.

The fuel pellets consists of coated TRISO fuel microspheres dispersed in a graphite matrix. As shown in Fig. 1, the fuel microsphere consists of either UC-NbC or UN-TaC fuel kernels, 500 µm in diameter, with a triple coating. The coatings inner layer is a low density pyrolytic graphite (PyC), 15-20 µm thick. The PyC coating accommodates the fission products recoil and partially accommodates the fission gases. An intermediate coating layer consists of high density graphite, 5-10 µm thick. This high density graphite strengthens the coating and suppresses the absorption of acids during the chemical Vapor Deposition process of the carbide outer coating. The outer layer is TaC or NbC coating, 10-20 µm thick. The outer coating is to protect against propellant interactions with the fuel. Although NbC is preferred over TaC because of its low neutron absorption cross section, its melting temperature is ~ 200 K lower than TaC.

The high melting temperature of the fuel material allows the peak fuel temperature in the core to reach as high as 3,100 K for a hydrogen exit temperature of 3,000 K. The core diameter is small, about 80 cm. Spacecraft shielding is divided with a hot shield within the pressure vessel cooled to between 600 and 680 K by the propellant and a shadow shield outside of the reactor in the cone of the propellant tanks. Both shields contain layered lithium hydride (LiH)-tungsten (W).

Liquid hydrogen is pumped from the tank through the nozzle torus to cool the nozzle structure, then flows axially through the Be_2C radial reflector where it transforms into a gas by the time it exits the reflector structure. The hydrogen gas then flows to the turbopumps and returns to the reactor dome where it flows downwards through the hot shield to maintain a temperature between 600 to 680 K in the LiH portion of the shield. The hydrogen flow from the shield, cools the upper axial Be_2C reflector, then flows downwards through the annular space between the core and the radial reflector. Flow continues radially through the core to remove the heat generated in the fuel pellets and exits through the central channel (~ 19 cm in diameter). The lower axial reflector is made of either BeO or Be_2C to sustain the high temperature at the exit of the core (3,000 K).

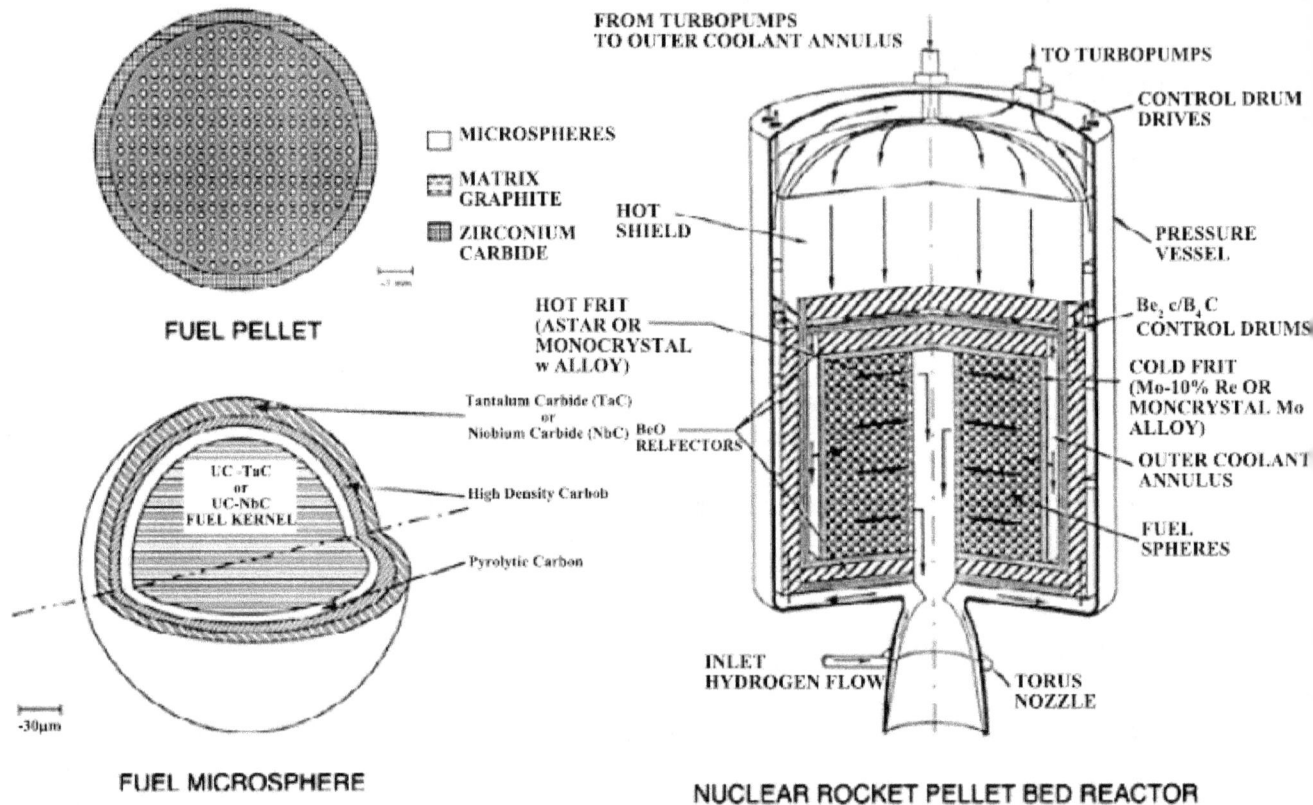

Fig. 1. Schematic of pellet bed reactor and fuel pellet. *From Mohamed S. El-Genk and Nicholas J. Morley, 1991.*

The radial flow arrangement significantly reduces the pressure losses in the core. This enables the operation at low system pressure (1 - 5 MPa). The hydrogen enters the core through a Mo-10% Re or a monocrystal Mo-alloy frit and exits through an ASTAR-811C or monocrystal W-alloy frit. The openings in these frits (< 1 cm in diameter) are sized to distribute the flow through the core commensurate with the axial power profile to limit the peak fuel temperature to less than 100 K above the gas exit temperature. For reactivity control, the core is surrounded with 26, 11-cm diameter, segmented B_4C/Be_2C control drums which are integrated into the radial reflector structure. Eight 4-cm diameter safety rods are also located 19 cm from the center of the core.

Neutronic analysis of the PeBR core showed the reactor design is not criticality limited. A high height-to-diameter (H/D > 1.85) reduces the pressure losses through the core and provides sufficient surface area for passive cooling of the reactor core in case of a loss-of-coolant accident and after operation reduces the mass of hydrogen propellant needed for decay heat removal. This effectively increases the average specific impulse by more than 4 percent. Also, neutronic analysis showed that either the control drums or the safety rods with a central B_4C plug (~ 9 cm in diameter) are sufficient to operate the reactor and keep it subcritical during a water immersion accident (k-effective < 0.94).

Representative design and operating conditions are given in Table 1. Thrust level is 315 kN, specific impulse is 1,000 s, peak fuel temperature is 3,100 K, and reactor specific mass is 1.0 kg / MW_t.

Table 1. Pellet Bead Reactor nominal core parameters for nuclear thermal propulsion.

Rated Reactor Power (MW_t)	1,500
Core Diameter (m)	0.8
Core Height (m)	1.3
Reactor Core Power Density (kW/cm^3)	3.0
Diameter of Central Coolant Channel (m)	0.2
Coolant	Hydrogen
Maximum Fuel Temperature (K)	3,100
Maximum Coolant Exit Temperature (K)	3,000
Core Inlet Temperature (K)	120
Reflector Inlet Temperature (K)	70-80
Coolant Flow Rate (kg/s)	32
Specific mass of reactor (excluding shield)[a] (kg/MW_t)	1.0
Fuel Type [melting point (K)]	UC-TaC (2,800-3.670[b] ± 50) UC-NbC (2,800-3.570[b] ± 50)

[a] Compared to about 1.8 kg/MW_t for a NERVA reactor.

[b] Melting point in equilibrium with carbon; single phase UC-TaC has a higher melting point.

The PeBR has some major potential advantages: potential specific impulse of 1,000 s, two independent control systems for redundancy in operation and safety during launch, passive removal of the decay heat from the core after operation, the absence of internal core support structure that eliminates internal structures that need to be cooled and allows fueling in space and facilitates end-of-life disposal. However, critical issues have not been address as part of an experimental program. These include nuclear fuels operating at peak temperatures of 3,100 K, materials compatibility issues, fuel coatings to suppress hydrogen erosion at high temperatures, high temperature refractory metal alloys with high strength and good compatibility with hydrogen, and fabrication of the spherical fuel pellets, Many of the material problems are common with other 1,000 s specific impulse concepts.

Wire Core Reactor Concept[3]

The wire core reactor incorporates a tungsten alloy wire fuel element. The core is constructed of layers of 0.8 mm diameter fueled tungsten wires wound over alternate layers of spacer wires. It is a derived of work originally performed in the aircraft nuclear propulsion program (710 Program). It's compact size is due to its high surface area per unit volume with about 19 cm^2 of heat transfer area per cubic cm of core compared to about 5cm^2 for a smooth passage reactor.

The active core is configured as annular in shape with radial coolant, see Fig.2. Hydrogen enters each end of the inlet duct, makes a 90 degree turn, flows through the active core, makes a second 90 degree turn, and then flows through the exit duct to the exhaust nozzle. The fuel consist of fueled tungsten clad wires 0.89 mm in diameter and unfueled tungsten wire spaces averaging 0.34 mm in diameter. The fuel is a cermet containing 70 vol % uranium nitride and 30 vol % tungsten. Cladding is a tungsten/rhenium alloy with 3 to 5% rhenium, varying in thickness from 0.076 mm at the inner core radius to 0.178 mm at the outer core radius.

To fabricate the tungsten wire fuel, braided tungsten tubing is used as a starting point. The braided tubing consist of 0.2 mm tungsten wire braided to a diameter of 3.2 mm. This braided tubing is then filled with 0.1 mm UN particles coated with tungsten and then chemical vapor deposited with tungsten. This tubing is swaged from the 3.2 mm diameter to 1.9 mm and drawn to the final wire fuel element of 0.89 mm diameter. A first layer of the wire fuel is wound on a mandrel with a spacing of about 0.7 mm. Across the fueled wire, spacer wires are placed about 13 mm. Additional layers of alternating fueled wires and spacer wires are added until the ring shaped portion of the core is completed. Each ring shaped portion is about 10 cm in the axial direction and 2.5 cm thick in the radial direction. The core is build up of 57 ring shaped segments that will be tungsten CVD to form a solid continuous core.

Design studies of 914 kN thrust engine were performed with a specific impulse of 1,000 s. The engine used an expander cycle as shown in Fig. 3 with heat generated in the nozzle and internal components to drive the hydrogen turbopump. Fig. 4 provides the dimensions for the rocket engine and Table 2 a mass break down.

Table 2. Reactor/shield assembly mass for 914 kN thrust engine.

Component	Mass (lb)
Active core	1,560
Reflectors	910
Gamma shield	1,060
Neutron shield	250
Control and actuators	190
Outer pressure shell	430
Inter pressure shell	250
Core rear support	110
Core front support	160
Core sheath	160
Total mass	**5,080**

The wire core reactor configuration has certain advantages with its large heat transfer area per unit volume, separation of fuel and structure, high power densities, no need for structural support in the heat gas flow direction and compatibility between uranium nitride, tungsten, and hydrogen even at elevated temperatures. However, only limited experimental work has been performed on this type of fuel form. Therefore, to reach its full potential, development is needed in fabrication, fueled wire performance evaluation and core assembly performance. This includes UN cermet fuel particle fabrication, braided wire fabrication, fueled wire swaging and drawing, wire winding and joining techniques, and sample fuel ring fabrication. Experimental verification is needed on $W-Re-UN-H_2$ compatibility at elevated temperatures, nitrogen overpressure in UN fuel matrix, fueled wire mechanical properties and in-core performance. Component, subassembly and assembly testing will eventually be needed.

Fuel – UN

Tungsten Cladding

Braided Tube Strands

Vapor Deposited Tungsten – Shaded Areas

Section A–A

Fuel Cermet

Typical Fuel Cermet

Braided Tube

Fuel Rod dia After Vapor Deposition

Fuel Rod Cross Section Prior to Extruding to Final Diameter

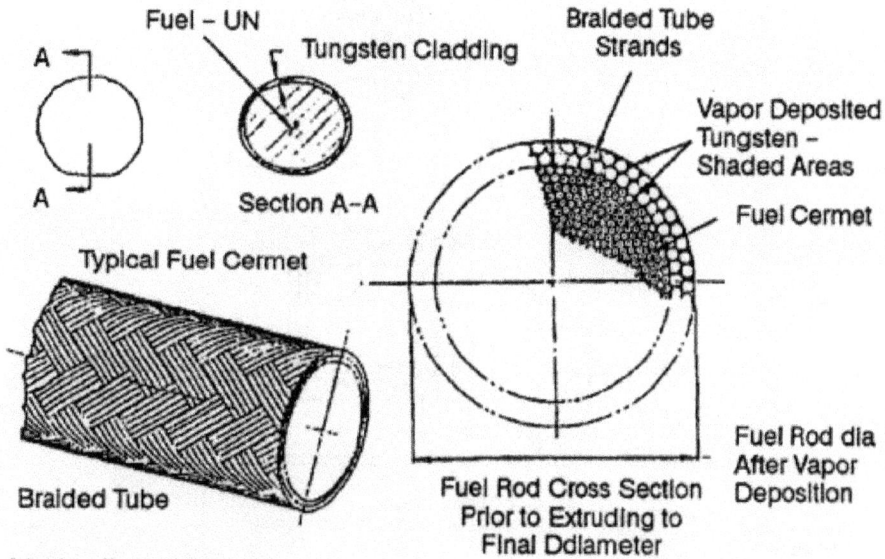

Making Fueled Wire

- Obtain 8–mil tungsten wire
- Braid wires into 3.2–mm–dia tubes
- Obtain 0.1–mm size UN fuel particles
- Coat fuel particles with tungsten

- Fill braided tube with coated fuel
- Vapor deposit tungsten on filled tube
- Swage tube from 3.2 mm to 1.9 mm
- Draw to 0.88–mil–dia finished wire

WIRE FUEL FABRICATION

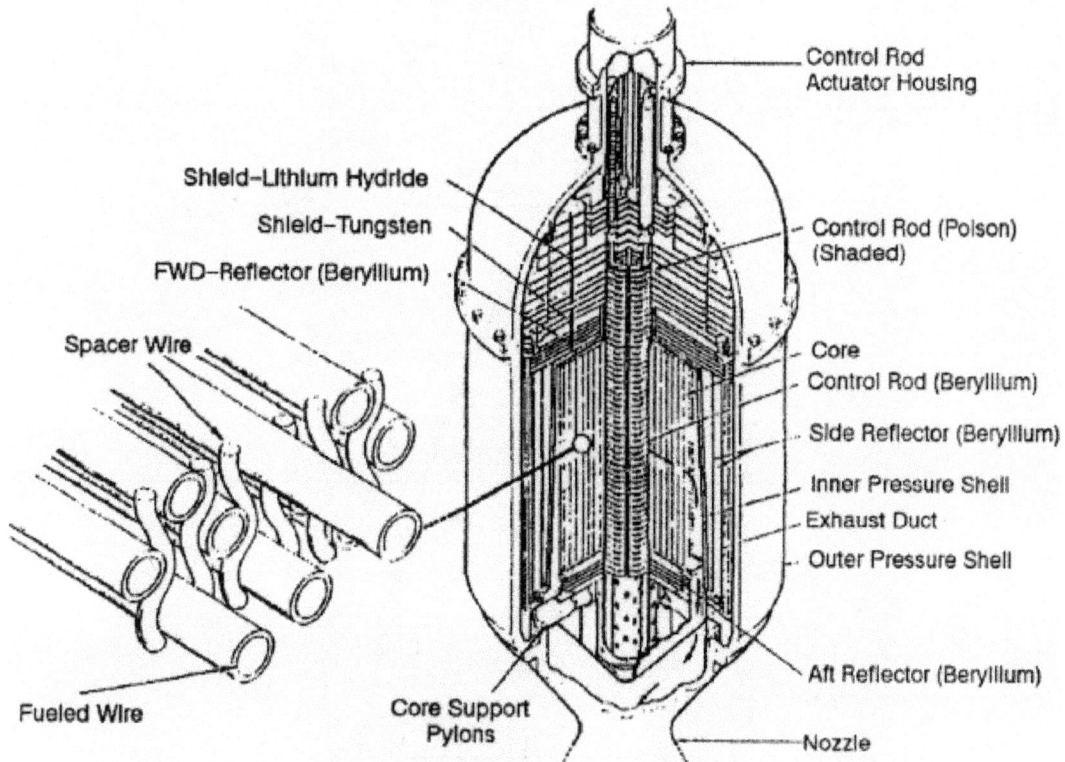

Control Rod Actuator Housing

Shield–Lithium Hydride

Shield–Tungsten

FWD–Reflector (Beryllium)

Control Rod (Poison) (Shaded)

Spacer Wire

Core

Control Rod (Beryllium)

Side Reflector (Beryllium)

Inner Pressure Shell

Exhaust Duct

Outer Pressure Shell

Fueled Wire

Core Support Pylons

Aft Reflector (Beryllium)

Nozzle

WIRE CORE REACTOR

Fig. 2. Schematic of wire core engine. *From Richard B. Harty and Robert G. Brengle, 1993.*

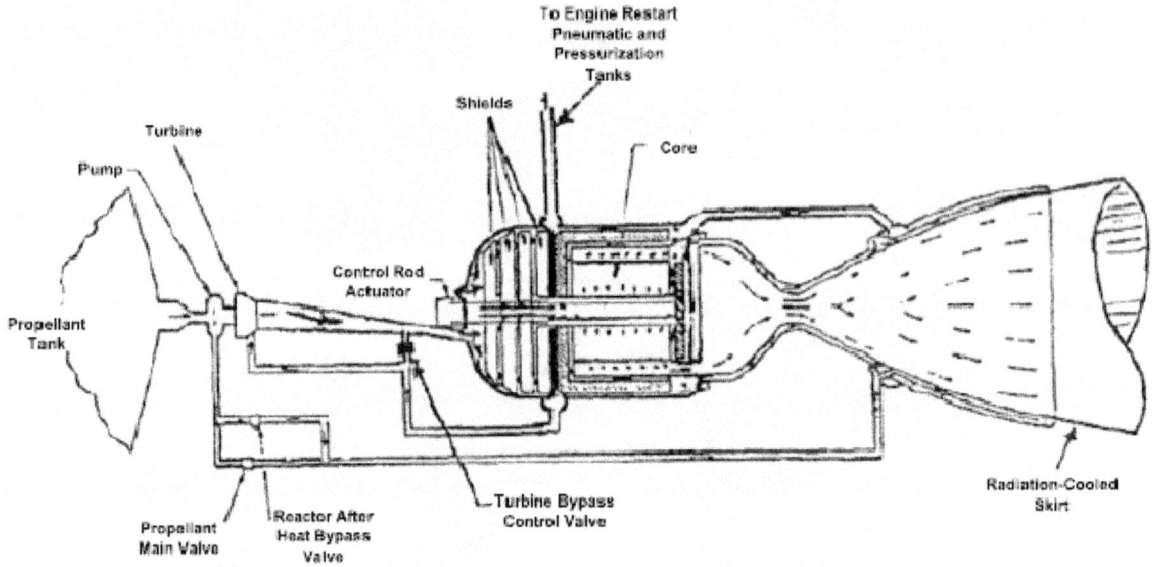

Fig. 3. Wire core reactor engine schematic using an expander cycle. *From Richard B. Harty and Robert G. Brengle, 1993*

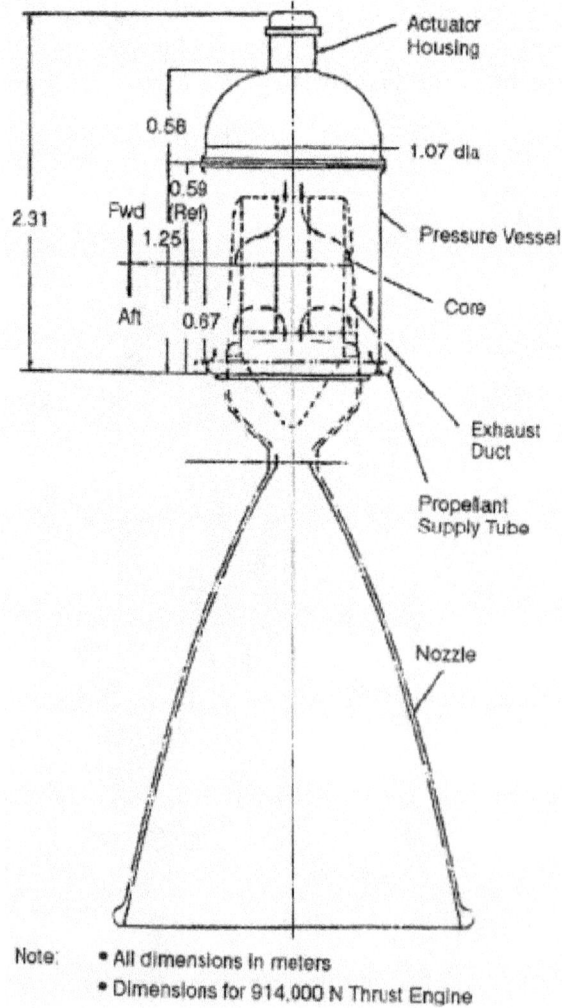

Fig. 4. Engine dimensions. *From Richard B. Harty and Robert G. Brengle, 1993.*

Conical Fuel Core[4, 5]

IMPULSE is a propulsion concept using a conical fuel geometry. The conical fuel element, shown in Fig. 5, consist of a thin fueled plate shaped as a truncated cone with a perimeter lip that is unfueled. The conical fuel elements are stacked together to form a fuel assembly. The element perimeter lip provides support and defines the spacing between adjacent elements; orificing placed in the lip regulates the coolant flow as it enters the fuel element; and each element can be orificed individually to match required coolant flow to core power distribution.

The conical fuel element can use a varied of fuels (see Table 3). Each of the fuels have different characteristic fuel loading and maximum operating temperatures. The carbide fuel has the highest temperature capability since it is basically unaffected by the hydrogen for temperatures up to 3,100 K, and a fuel loading fraction capability of 0.2. The cermets also have a good resistance to corrosion and are useable for temperatures up to 2,800 K. They can accommodate fuel fractions of 0.5 or higher, which makes it very effective for small cores.

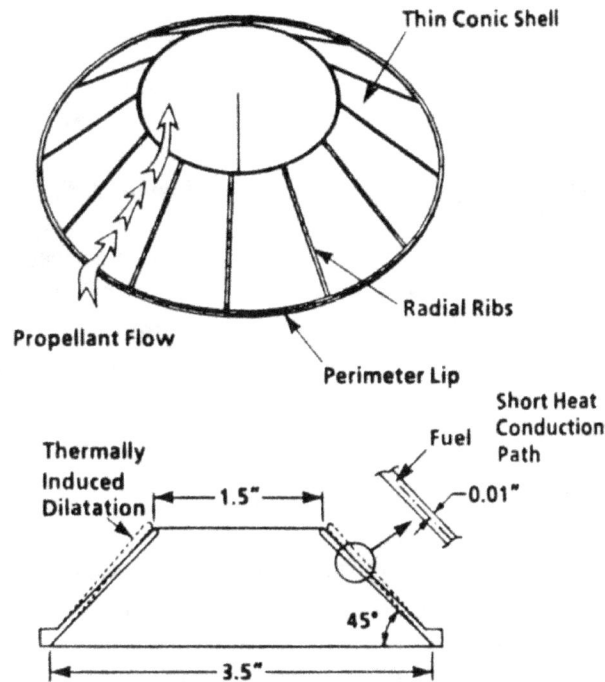

Fig. 5. Conical fuel element. *From Lyman J. Petrosky, et al., 1993.*

Table 3. Specific impulse fuel performance characteristics.

Fuel	Maximum Temperature (K)	Maximum Fuel Fraction	Specific Impulse (seconds)
Carbide	3,000	0.2	985
Composite	2,700	0.2	900
Moly-Cermet	2,200	0.5	805
Tungsten-Cermet	2,800	0.5	920

The reference configuration for IMPULSE is a 19 fuel assembly, ZrH moderated thermal reactor with a three pass core (see Fig. 6). This allows the use of a heterogeneous core of different fuels for the low and high temperature region of the core. The reference design for 334 kN thrust (75,000 lb$_f$), and with carbide fuel cones delivers an specific impulse of 970 s with a thrust/weight ratio of 30.

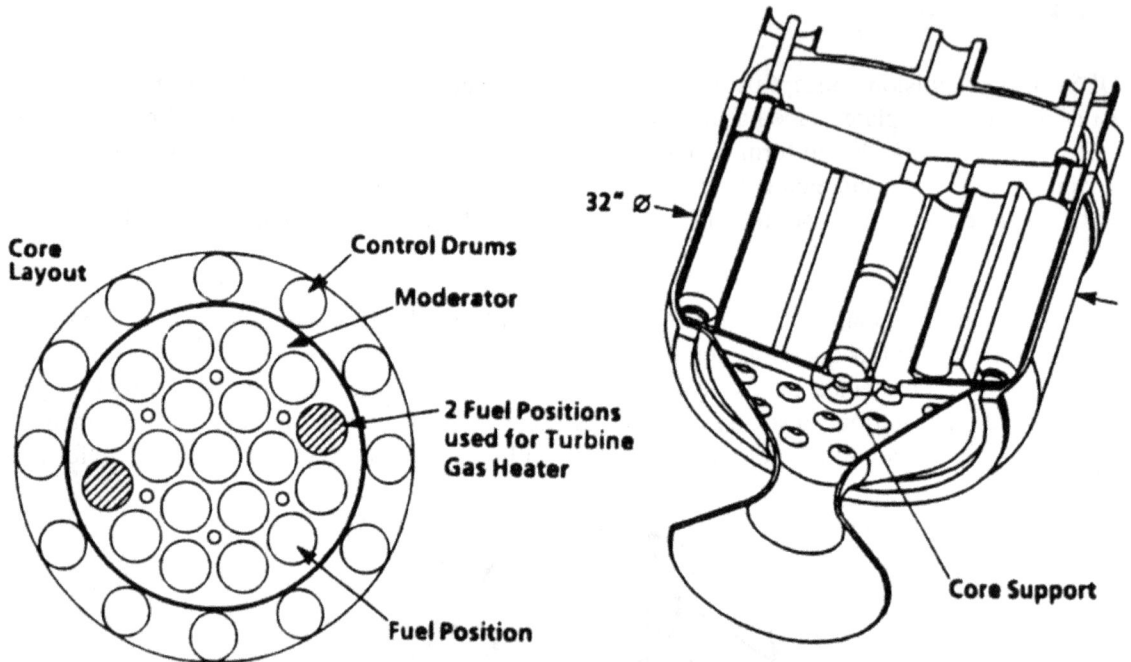

Fig. 6. IMPULSE core arrangement. *From Lyman J. Petrosky, et al., 1993*

Advantages of this configuration include: coolant flow paths are well defined with no outlet flow restrictions; thermal gradients across fuel thickness and flow channels are nominal, resulting in low stresses; truncated conical shape reduces thermal stresses and assures constant flow paths; and fuel arrangement provides mechanical stability in core. However, though the concept is based on existing materials, no experimental work has been performed.

Pressure Fed Nuclear Thermal Rocket (PFNTR)

The Pressure Fed Nuclear Thermal Rocket (PFNTR) is a low pressure engine that utilizes conventional propellant tank pressure for expelling the liquid hydrogen from the tank through the flow control system to the reactor, without the use of a turbopump. The advantages of such a system are simplicity and the advantage of increased specific impulse with dissociation of hydrogen that occurs at operations below 1.0 MPa and chamber temperature above 3,000 K.[6]

Various configurations have been proposed with different fuel configurations and flow paths. The radial outward flow of hydrogen concept, shown in Fig. 7, features a spherical core to maximize the high temperature heat transfer area. Fuel can be particles, pellets or platelets. Reactivity in the core is controlled by the amount of hydrogen in the core, thereby eliminating control drums. Operating at a low pressure of 0.1 MPa (15 psia) is projected to augment the heat transfer as a result of hydrogen dissociation-recombination effects.[7,8]

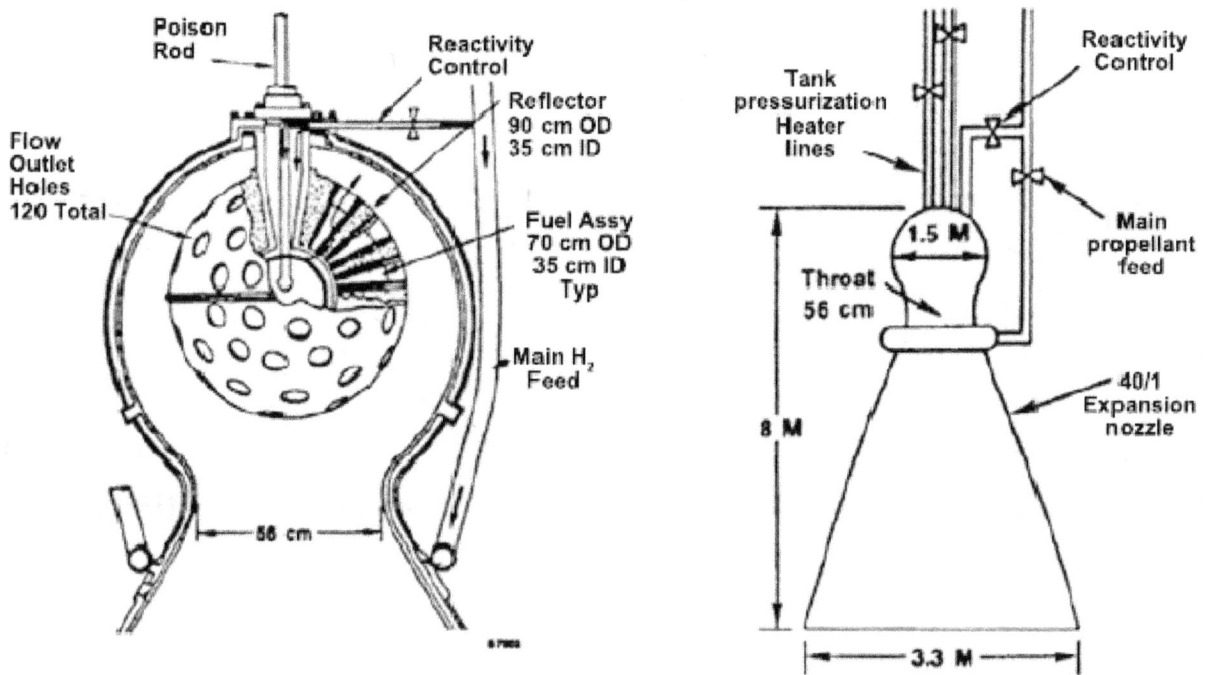

Fig. 7.Schematic of Low Pressure Nuclear Thermal Rocket. *From Carl F. Leyse, et al., 1990.*

The benefits of this concept depend mainly on the dissociation of hydrogen at high temperatures. Fig. 8 shows the theoretical gains in specific impulse. However, the delivered performance calculated by applying kinetic, divergence and boundary-layer loss terms to the shifting equilibrium results using a nozzle expansion ratio of 500 shows much reduced benefits (see Fig. 9). A comparison of ideal and delivered performance is shown in Table 4.[9]

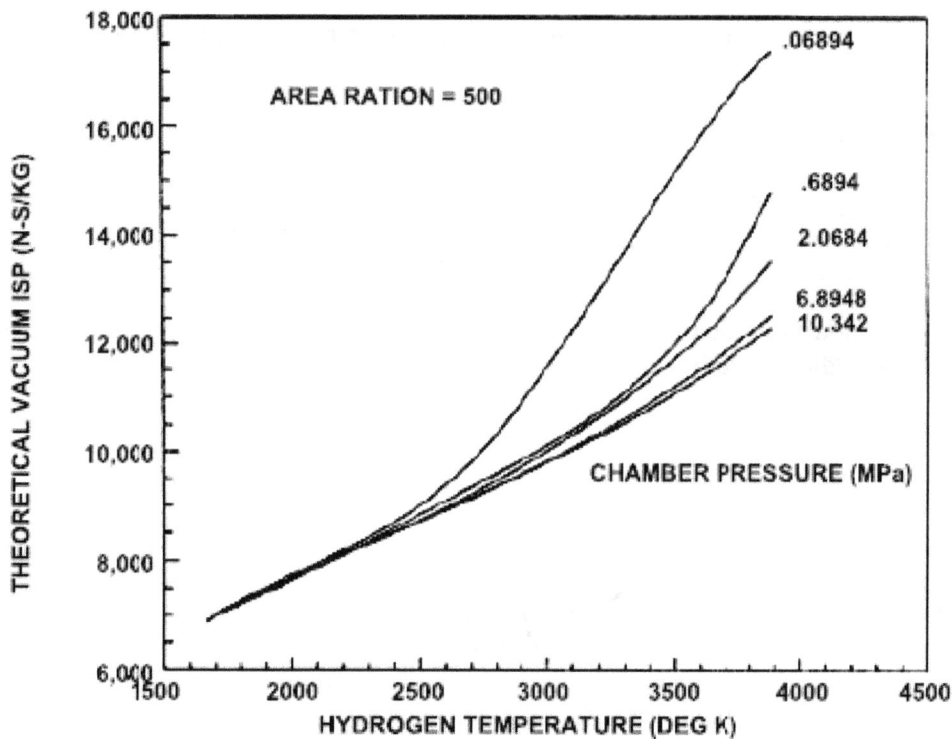

Fig. 8. Theoretical specific impulse versus temperature for a nozzle area of 500:1. *From J. Wesley Davis, et al., 1991.*

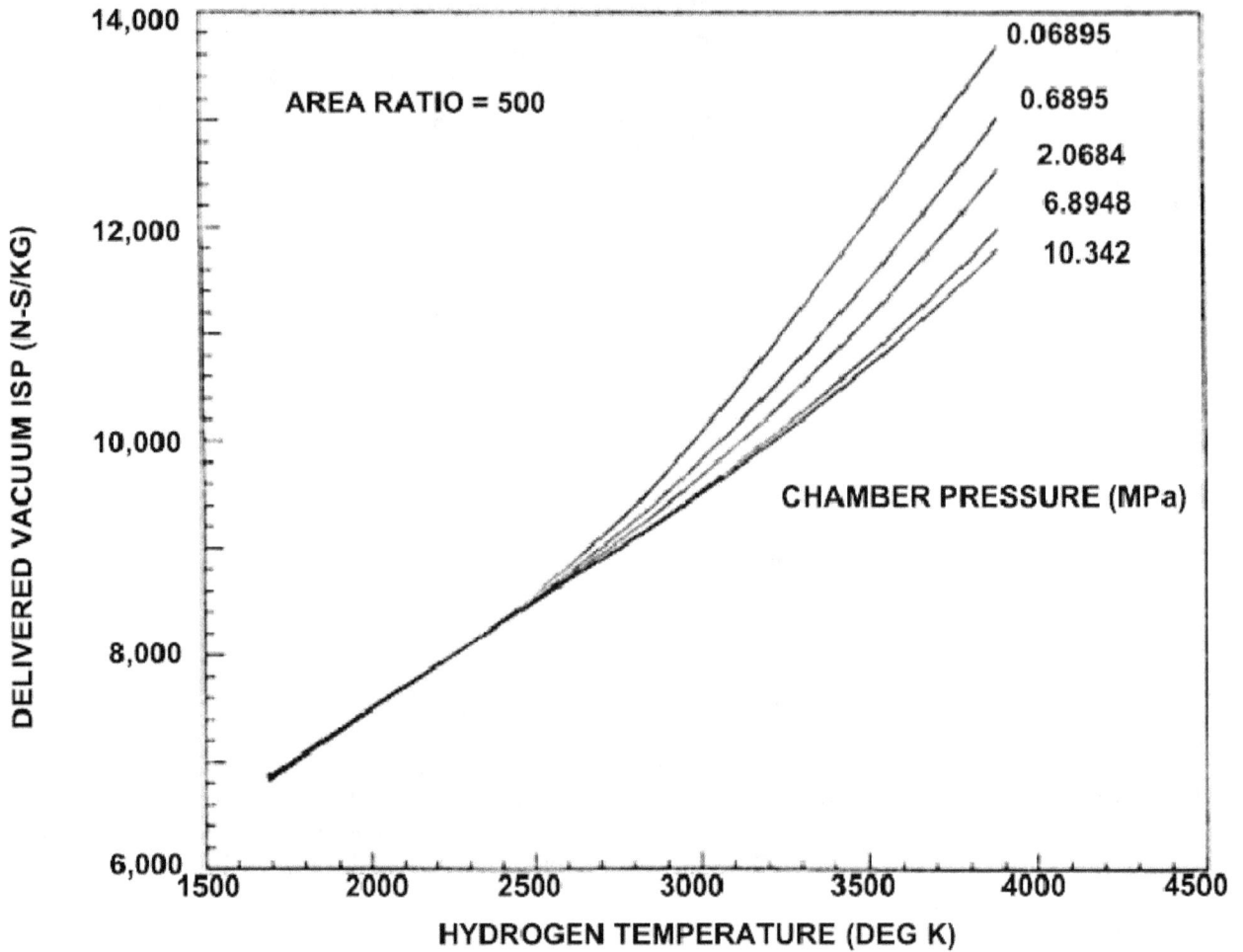

Fig. 9. Delivered specific impulse versus temperature for a nozzle area of 500:1. *From J. Wesley Davis, et al., 1991.*

Table 4. Ideal versus delivered performance.

	Ideal Isp (N-s/kg)		Delivered Isp	
	3,100 K	**3,500 K**	**3,100 K**	**3,500 K**
68.95 kPa	12,200	15,000	10,430	12,000
10.34 MPa	9.900	11,000	9,600	10,800

Experimental verification is needed of ideal versus delivered specific impulse at high temperatures. Also, reactor control and shielding could be major issues.

Liquid Core Concepts[10]

The Liquid Annulus Reactor System (LARS) and the DROPLET CORE concepts increase the fuel temperature to the liquid state and are projected to operate between 3,000 - 9,000 K. The LARS fuel element consists of a rotating cylindrical can that holds an inner layer of a high temperature refractory material. This refractory material contains uranium and an appropriate diluents(s), possibly a mixture of UC_2 and ZrC. The thin outer layer adjacent to the fuel element can of the refractory fuel is solid while the inner layer adjacent to the flowing hydrogen is liquid. The liquid refractory is maintained as an annular layer by rotating the fuel element can at a speed sufficient to stabilize the molten fuel layer. Fig 10 and 11 illustrate the concept.[11]

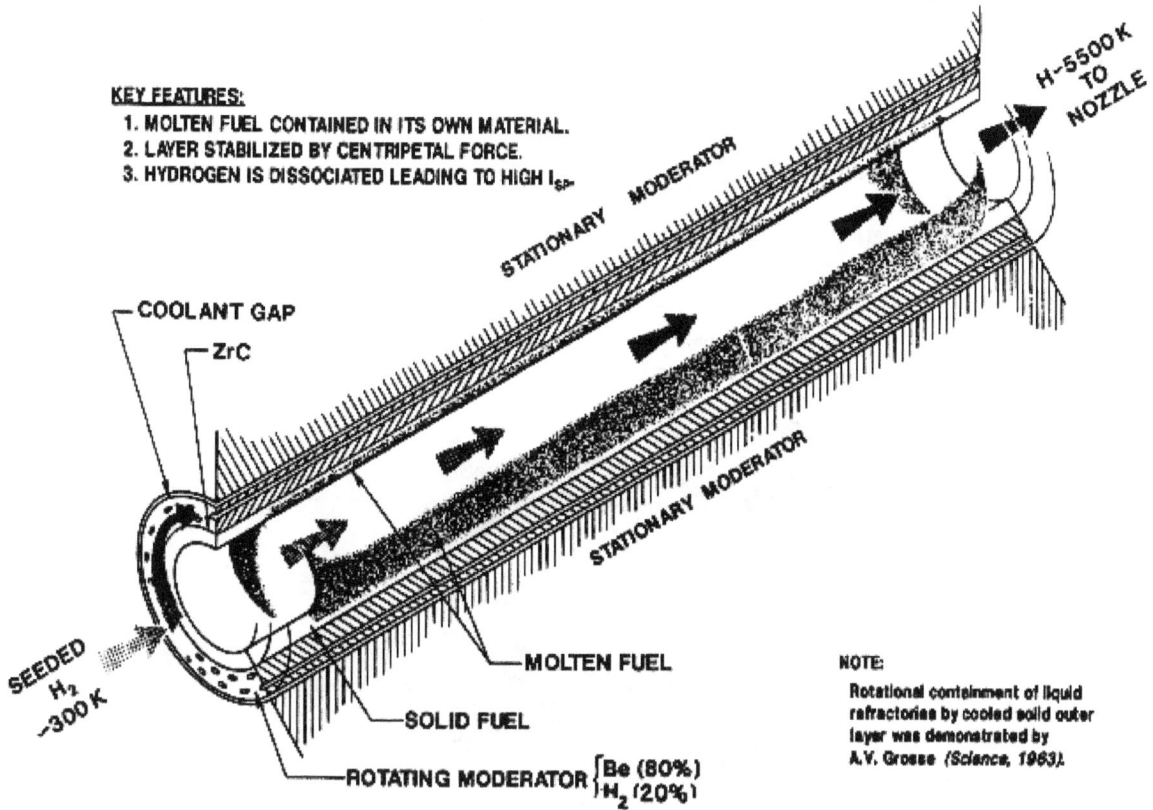

KEY FEATURES:
1. MOLTEN FUEL CONTAINED IN ITS OWN MATERIAL.
2. LAYER STABILIZED BY CENTRIPETAL FORCE.
3. HYDROGEN IS DISSOCIATED LEADING TO HIGH I_{sp}.

NOTE:

Rotational containment of liquid refractories by cooled solid outer layer was demonstrated by A.V. Grosse *(Science, 1963).*

Fig.10 LARS concept.

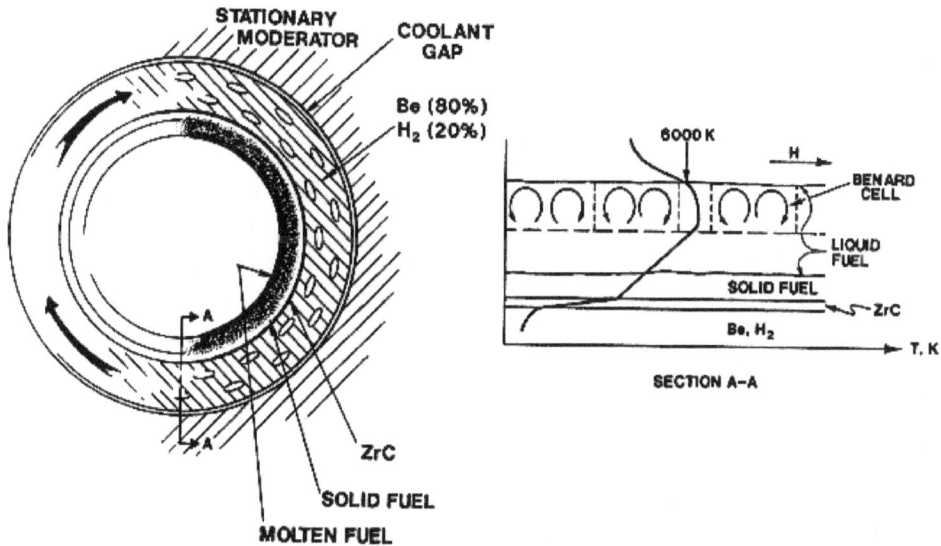

Fig. 11. LARS rotating fuel element.

The concept has major technical issues including the stability of the liquid layer, materials, evaporative loss of fuel, radiative and transport properties of liquid fuel, and compatibility with hydrogen, heat transfer, and physics questions related to high temperature cross sections.

The DROPLET CORE concept, was conceived to eliminate strength requirements on fueled materials at high temperature. Fueled particles such as W-coated UO_2 are suspended in flowing hydrogen. A body force prevents the fuel particles from flowing out with the hydrogen, see Fig. 12.[12] In space, the body force can be provided by the acceleration of the rocket, the rotation of the reactor, or by a hydrodynamic force such as vortex flow within the reactor. Rotation or vortex flow allows the hydrogen to flow through the fission heated articles while they are centrifuged toward the walls. The hydrogen is heated by direct contact with the fuel particles. Potential problems with this concept include expected fission product release during normal operations and the agglomeration or sintering together of the fuel into clumps. No experimental work has been performed to support this concept.[13]

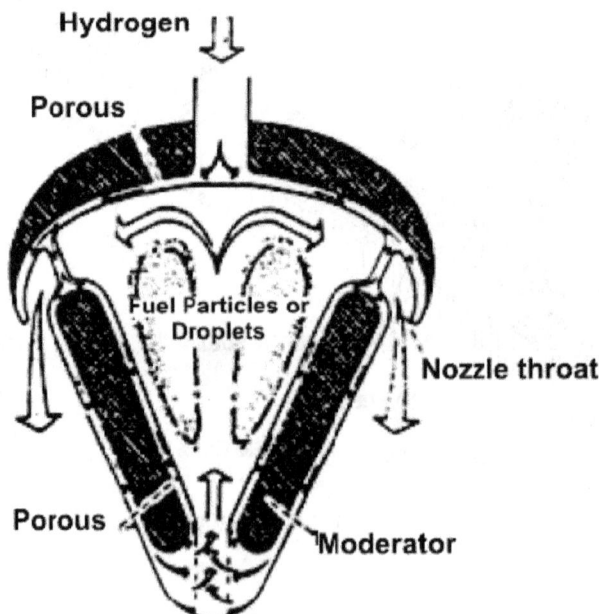

Fig. 12. Solid particle or liquid droplet reactor. Gaseous Core Reactors

Gaseous Core Reactors

The interest in gaseous core reactors have historically been as a space propulsion application. Large performance improvements over solid-core nuclear rockets are postulated with specific impulse gains over solid core fission reactors possibly by a factor of seven. Specific impulse values up to 50,000 m / s and nuclear fuel temperatures in excess of 10,000 K are postulated.[14] The uranium fuel in such systems would be in the plasma state The gaseous core reactor concept has an externally moderated cavity assembly containing the nuclear fuel in the gaseous phase. For system temperatures above about 5,000 K (a temperature regime ideal for nuclear rocket applications), vaporized uranium metal would serve as the nuclear fuel in the form of a fissionable plasma. The gaseous nuclear fuel is separated from the cavity walls by hydrodynamic techniques, making it possible to operate systems at temperatures exceeding the structural limit temperatures of their containment materials.

Many gaseous core concepts were examined, with the coaxial flow system and the nuclear lightbulb system emerging as the favored approaches (see Fig. 13).[15]

The coaxial flow concept involves a low velocity inner stream of fissioning uranium metal plasma, surrounded by a very high velocity propellant gas (typically hydrogen) stream. The hot inner fuel plasma transfers thermal energy to the outer propellant stream by both convection and radiation heat transport. In this configuration, heat transfer occurs essentially unimpeded. However, a difficulty is that the propellant flow stream can also physically mix with the uranium plasma, leading to a loss of nuclear fuel.

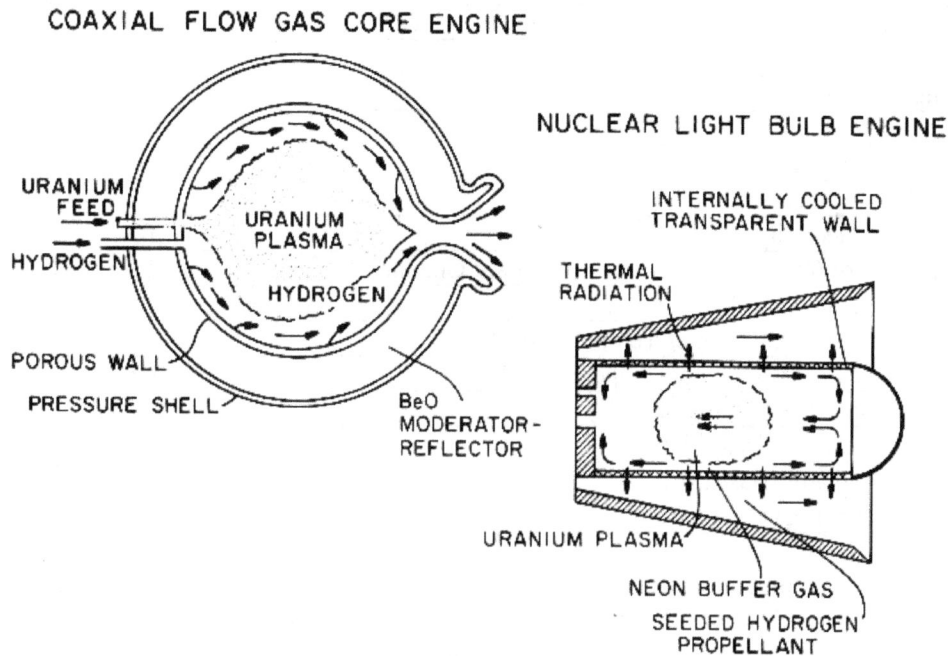

Fig. 13 Plasma core rocket concepts, including coaxial flow gas core engine (left) and nuclear light bulb engine (right). *From K. Thorn and F. C. Schwenk, 1997.*

In the nuclear lightbulb concept, the fissioning plasma is confined within a transparent (quartz) cell and is kept away from the cell's walls by the swirling flow of a tangentially injected buffer gas (e.g., argon). Here, energy transfer from the plasma to the propellant is by radiation heat transfer alone; the two streams do not physically mix. Even though the quartz cell effectively confines the nuclear fuel, radiation heat transfer is limited to selected regions of the electromagnetic spectrum. This limit is established by the transparency of the cell material and its resistance to nuclear radiation induced "darkening".[16, 17]

Both of these gaseous core reactor concepts share the common engineering problem of promoting effective radiation heat transfer to the flowing propellant, while avoiding thermally induced destruction of the cavity materials. Since neutrons and gamma rays from the fissioning plasma may not be totally absorbed by the propellant stream, nuclear radiation heating of the cavity walls must also be effectively handled. Representative preliminary performance estimates for a nuclear lightbulb engine with a cold beryllium reflector is given in Table 5.

Table 5. Preliminary performance estimates for nuclear lightbulb rocket engine with cold beryllium reflectors.

Cavity Power (MW$_t$)	189
Power to Propellant (MW$_t$)	168
Reflective Liner Heat Load (MW$_t$)	21
Be Reflector Heat Load (MW$_t$)	7.5
Specific Impulse (m/s)	18,000
Number of Cavities	One
Thrust (N)	12,500
Engine Mass (kg)	4,620
Moderator (kg)	1,860
Pressure Vessel (kg)	1,830
Nozzles, Pumps, etc., (kg)	930
Cavity Length (m)	1.83
Cavity Diameter (m)	0.49

Cavity Volume (m^3)	0,34
System Pressure (MPa)	50.7
Fuel Material	^{233}U plasma
Buffer Vortex Fluid	Neon
Reflector Material	Be
Reflector Liner Material	BeO
Liner Reflectivity	0.95
Fraction of Cavity Surface with Propellant Channels	0.25
Propellant Mass Flow (kg/s)	0.68
Beryllium Coolant Outlet Temperature (K)	610
Effective Fuel Radiating Temperature (K)	8,300
Propellant Exit Temperature (K)	6,700
Cavity Power Density (MW$_t$)	566
Space Radiator Mass (kg)	2,380
Total Mass (kg)	7,000

Another gaseous core reactor concept involves the use of a "mini-cavity," surrounded by a BeO moderator region and a nuclear-fueled "driver" region (see Fig. 14).[18] The uranium carbide fuel element driver region provides the majority of the neutrons needed for maintaining the chain reaction and substantially reduces the criticality dimensions of the gaseous cavity region.

Analytical and experimental studies were performed for the coaxial flow concept.[19] These efforts examined hydrodynamic behavior and determined the rate at which nuclear fuel needed to be fed into a cavity to compensate for propellant stream losses. Experiments included demonstrating vortex confinement of uranium hexafluoride gas at densities high enough to sustain nuclear criticality. Also, techniques for injecting UF$_6$ into confined vortex flows were demonstrated. Through both theory and experiments, it was shown in order to establish laminar flow at the fuel plasma that the tangential velocity at the cavity wall must be significantly greater than the axial velocity of the buffer gas. Vortex injection areas on the order of 90 cm^2 appeared necessary to produce a flow pattern with good confinement characteristics and laminar recirculation cells. The corresponding tangential injection velocity for the argon buffer gas was 360 cm / s, or about 18 times the maximum axial velocity of 20 cm / s.[20]

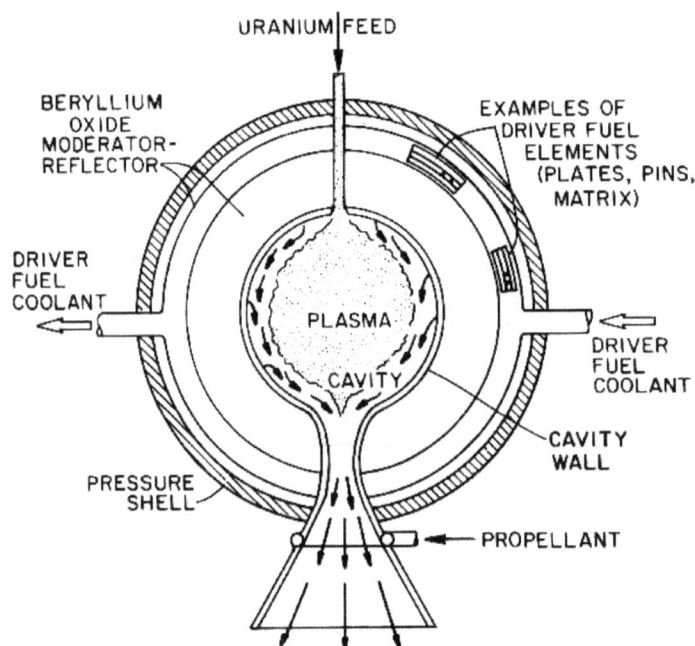

Fig. 14. Minicavity plasma core rocket engine. *From* Stanley Chow, 1976.

Basic research on gaseous core reactors demonstrating reactor physics and radiation heat transfer principles has been completed. These efforts included a series of static UF_6 reactor experiments at temperatures between 300 and 500 K and power levels of 0.1 to 1.0 kW. These were followed by flowing UF_6 reactor physics experiments and flowing argon buffer gas physics experiments, which successfully demonstrated the confinement of a fissioning U(93%)F_6 plasma inside an argon vortex. For example, on 10 May 1979 for a UF_6 flow rate of 1.8 g / s, a power level of 30 kW was sustained for approximately 120 seconds with excellent reactivity stability. Reactivity measurements indicated the achievement of a confined nuclear fuel mass of 35 g of UF_6 and a corresponding vortex residence time of 19 seconds. Similarly, for a nuclear fuel flow rate of 2.7 g / s, a power level of 10 kW was maintained for 140 seconds. In this case, a confined nuclear fuel mass of 50 g corresponded to a residence time of 18 s, indicating (as anticipated) that the confinement mechanism was independent of the flow rate.[21] The next logical technical step in the development of a gaseous core reactor system would be a reactor feasibility demonstration experiment. There are, however, no current plans for such a demonstration. The key concern with the open cycle gas core nuclear engine remains stability of operation.[22]

In 1983, Diaz and Dugan proposed a Nuclear Vapor Thermal Rocket Engine (NVTR) that isolates the gaseous fuel (either UF_6 or UF_4) from the hydrogen propellant.[23] The NVTR shown in Fig. 15[24] uses a modified-NERVA geometry and systems with the solid fuel replaced by highly enriched (> 85%) uranium tetrafluoride (UF_4) vapor. The cell geometry is modified to meet neutronic and heat transfer requirements. The NVTR uses a graphite moderated reactor and hydrogen propellant is used to regeneratively cool the structure and moderator. The NVTR core is surrounded by BeO and C-C composite reflector regions. The NVTR is different from most gas cooled reactors in that most of the neutron moderation occurs within the core rather than in the external reflector.

Exit temperatures are 3,100 to 3,400 K (as compared to 5,000 K for the gaseous core concepts discussed above) with specific impulse in the 1,000 to 1,200 second range. Hydrogen coolant/propellant is maintained at high pressures on the order of 10×10^6 pascals. The hydrogen propellant also provides some moderation in the NVTR core though the majority of the moderation is provided by the graphite. The fuel element or fuel cell shows the ZrC or C-C composite forming channels that carry the hydrogen propellant. ZrC is compatible with hydrogen and the coolant channels in this system will not require a protective coating. However, the C-C composite will require a liner or protective adhesive coating for the coolant channels. Possibly the best liner material is ^{180}HfC. The ^{180}Hf has a thermal neutron capture cross section that is about eight times lower than that of natural hafnium; natural hafnium contains about 35% by weight ^{180}Hf.

The advantage of this type of core is that the fuel is relatively inexpensive, readily available, and a simple fuel form. Thrust-to-weight ratios are of 3 to 5 compared to 1 to 2 for large ZrC-moderated reactors. Ultra high temperature materials to isolate the propellant from the fuel become the driving factor in the design.

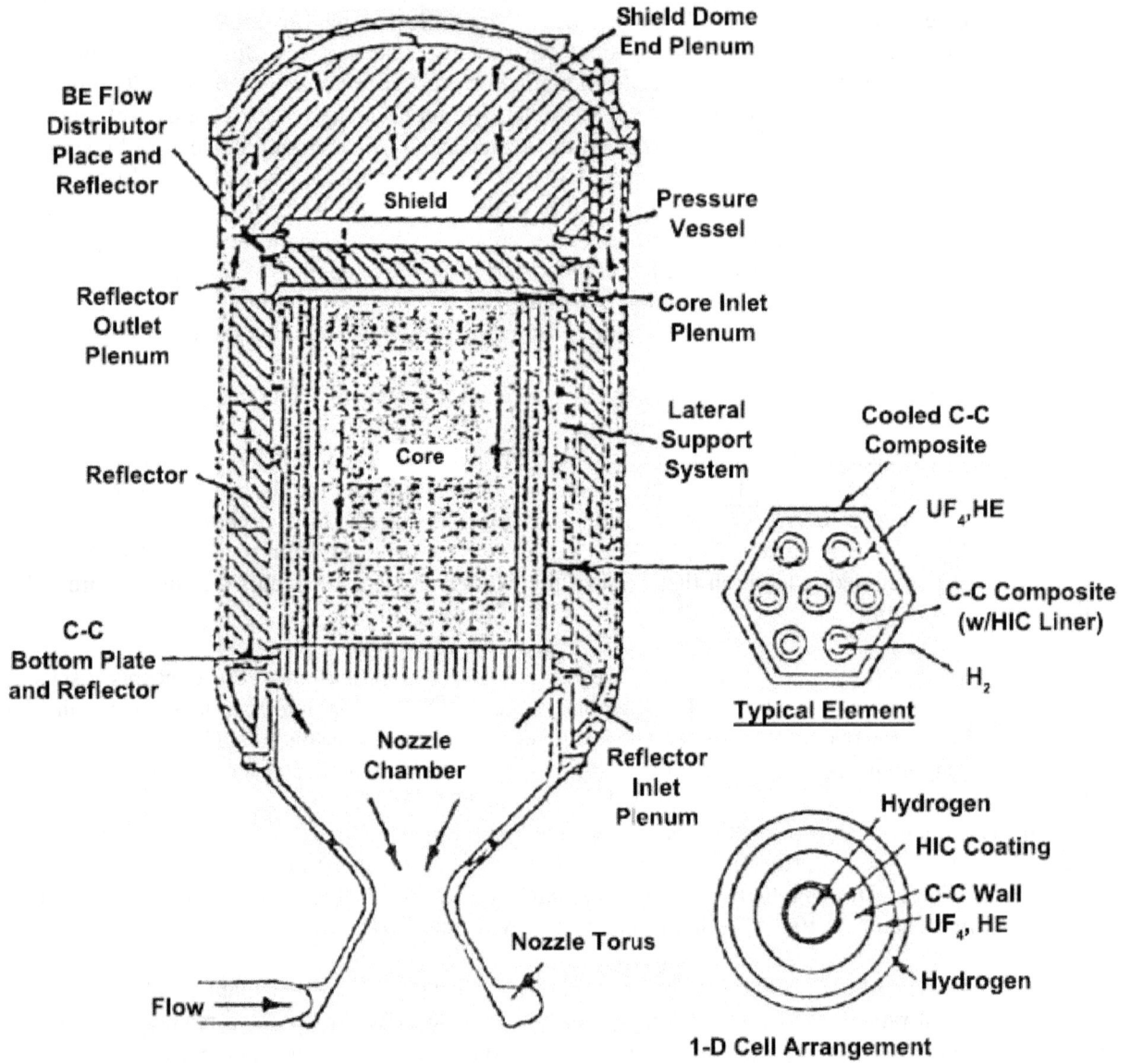

Fig. 15. Conceptual Nuclear Vapor Thermal Rocket engine. *From Nils J. Diaz, et al., 1991*

Chapter 7

NTP Key Development Issues

A number of key topical issues will be reviewed in this chapter. These issues are common to whatever nuclear thermal propulsion concept is eventually developed. Because the NERVA program has demonstrated components in the engine system outside the reactor core, we will concentrate our discussion on NTP fuels for solid core systems. Solid core reactors are the most likely first generation of NTPs and the same fuel materials are common to many designs. Next, we will discuss safety issues relevant to the use of nuclear thermal propulsion. This is followed by a discussion of engine cycle considerations. Then, we will address mission success and crew safety relevant to NTP using a manned Mars mission to illustrate the issues and solutions. A section will cover testing of NTP designs to satisfy revised, more restrictive radiological environmental standards. Finally, we will discuss the overall relative maturity of NTP concepts.

NTP Fuels[1]

The heart of any nuclear thermal propulsion systems is the fuel. Desirable characteristics of nuclear fuels include high density, high thermal conductivity, high melting point, high temperature stability, chemical compatibility, predictable irradiation performance, and ease of fabrication. Table 1 summarizes properties and characteristics of oxide, nitride, and carbide fuels. Carbide fuel performance, melting point, stability, fabricability, and compatibility favors them for NTP applications. Uranium oxide, in the form of tungsten metal matrix cermets, was developed for nuclear rocket fuels during the early 1960's. Although cermet fuels are robust, compatible with hot hydrogen, and resistant to fission product release, development was discontinued in favor of carbide fuels for the Rover program. Other U.S. NTP designs are based on Rover technology data.

Table 1. Characteristics of space reactor fuels.

Characteristics	UO_2	UN	UC	UC_2	$(U_{0.2}Zr_{0.8})C_{0.99}$
U Density, g/cc	9.66	13.52	12.97	10.60	2.88
Melt Point, K	3,100	3,035	2,775	2,710	3,350
Thermal Conductivity, W/mK	3.5	25	23	18	30
Relative Stability	Moderate	Low	High	High	High
Relative Swelling	Low	Mid	Mid	Low	High
Fission Gas Release	High	Low	Mid	Low	Mid
Fabricability	Ease	Moderate	Easy	Difficult	Difficult

Demonstration of stable high-temperature carbide fuels was the key to the success of the Rover/ NERVA programs. Three types of fuels were developed: (1) pyrolytic carbon-coated UC_2 spheres, dispersed in a graphite matrix and coated with NbC; (2) composite graphite/uranium, zirconium carbide fuel coated with NbC or ZrC; and (3) single-phase uranium, zirconium carbide $((U,Zr)C_x)$ fuel. Fig. 1 summarizes fuels information including a schematic representation of the macrostructure of the three Rover fuel types. The potential temperature capabilities are indicated with arrows on Fig. 2.

Type	Bead Loaded Graphite	Carbide-Graphite Composite	Carbide
Development Period	1958 -1967	1967 - 1972	1970 - 1972
Fuel Element			
Macrostructure			
Scale reference	20 vol. % particles; dia. = 125 ± 25 μm	35 vol. % carbide; carbide filament cross section 5-20 μm.	Mean grain size, 15 μm.
Fuel Composition	Pyrographite Coated UC_2 Beads	$(U_{0.085}Zr_{0.915})C_{0.98}$ + C Composite	$(U_{0.025}Zr_{0.975})C_{0.958}$ Carbide
Density, g/cm^3	2.30	3.50	5.50
U Loading, g/cm^3	0.400	0.400	0.300
Melting Temperature, K	2725	2900	3200
CTE (300 to 2400 K), μm-/m•K	5.8	6.7	7.8
Thermal Conductivity (300 K), W/m•K	110	80	7
Flexure strength (300K), MPa	35	55	76
Peak Fuel Temperature, K	2600	2525	2525
Exit Gas Temperature, K	2550	2450	2450
Carbon Loss Rate by Hydrogen Corrosion, g/cm^2-s			
NbC Protective Coating	$11.2 \cdot 10^{-6}$	--	--
ZrC Protective Coating	$5.5 \cdot 10^{-6}$	$2.8 \cdot 10^{-6}$	--
Carbide Fuel Form	--	--	$0.6 \cdot 10^{-6}$ [a]

[a]Estimated for fuel temperature of 2525 K from data of McMillan (1991).

Fig. 1. Evolution of Rover fuel types. *From* R. Bruce Matthews, et al, 1994.

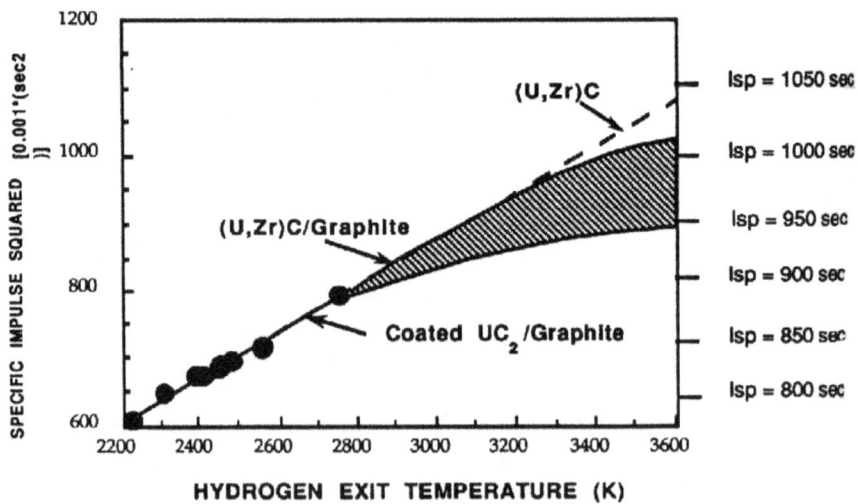

Fig. 2. Potential specific impulse as a function of hydrogen exit temperatures. *From* R. Bruce Matthews, et al. 1994.

Pyrolytic Carbon-Coated UC_2 Spheres were fabricated by extrusion to produce tens of thousands of fuel rods with both hexagonal and cylindrical cross-sections. Dispersed fuel was the most highly developed type. Early dispersed fuels were UO_2 particles mixed with graphite and a binder. The mix was homogenized and extruded into the typical Rover nineteen hole fuel rods. The fuel rods were baked to remove volatiles and heat treated to convert the oxide to carbide fuel and complete graphitization.

Mid-band corrosion problems led to the development of (U,Zr)C + graphite composite fuel. Because both components were continuous and interconnected, the composite fuel rods were more robust and chemically stable then the bead-loaded fuel rods.

Single phase (U,Zr)C was developed to eliminate the carbide/carbon eutectic limitation and permit higher operating temperatures. Program termination limited the number of carbide rods to seven tested in the nuclear furnace and therefore insufficient data was generated to fully evaluate performance potential. The solid solution is difficult to fabricate, tends to crack, and shows some corrosion in flowing hydrogen. However, singe-phase carbide fuel has the greatest potential for improving the performance of the next generation of nuclear rockets.

To protect the graphite fuel rod surfaces and coolant channels from hot hydrogen corrosion, chemical-vapor-deposited NbC and ZrC coatings were developed. In general these coatings exceeded expectations. Pyrocarbon coatings were used to protect the UC_2 particles from air oxidation; the mixed carbide ternary fuel compounds are stable in air and need no such protection. NbC and ZrC coatings were later used to prevent hydrogen reaction with the graphite matrix. These were not originally intended to be a fission product barrier, but did turn out to act as a fission product diffusion barrier. In PEWEE-1, pyrolytic carbon coated UC_2 particles dispersed in ZrC-coated graphite operate at 2,555 K for 40 minutes. Operational limits were the carbon/UC_2 eutectic limited operating temperature to 2,725 K, and "mid-band" corrosion cracking caused by thermal expansion mismatch between the graphite matrix and ZrC coating. The cracking in the ZrC coating caused reaction between hot H_2 and graphite that resulted in measurable mass loss from the fuel element.

The most recent work on carbide fuel fabrication focused on the high temperature mixed carbides. The properties which make the ternary carbide fuel compositions so desirable as a reactor fuel also makes their fabrication difficult. Methods that have been used to produce refractory carbide nuclear fuels include extrusion hot pressing, arc melting, cold pressing and sintering, sol gel, combustion synthesis, and freeze drying. Extrusion has the advantage that large quantities of material can be rapidly processed, and is viewed as a relatively low risk, highly proven method for producing refractory carbide fuel rods. This was demonstrated near the end of the Rover program. The lack of significant quantities of graphite in the mix required high extrusion pressures. Arc melting produces very dense parts with excellent mechanical properties. However, it is a relatively slow processing technique and thus is best for making laboratory test specimens. Cold pressing and sintering are relatively low risk methods for producing refractory carbide materials of variable fuel geometries, such as pellets and short rods. Combustion synthesis at high reaction temperatures requires relatively simple equipment setup, it is quick, it is energy efficient, and it can produce relatively pure final product. The formation reactions of mixed carbides from their elemental components are highly exothermic and self propagating. The high melting point results in the product retaining its original shape. UC_2 gel-spheres, using a modified process, the process quickly freezes sprayed droplets of aqueous suspensions of solids to form frozen spheres. The frozen spheres, containing UO_2, ZrC, and carbon, are freeze dried to remove water, then converted to the carbide, and sintered to density.

To protect the graphite from reacting with the hydrogen propellant, chemical-vapor-deposited refractory metal carbide coatings are applied to graphite matrix fuel rods and spheres. Also, the graphite matrix fuel rods are coated with ZrC. The chemical-vapor deposition techniques for the refractory carbide coatings are generally well know.

Carbide fuels activities have focused on increasing operating temperature, lifetime, and margins to failure. The major performance limitations are: (1) melting point of the fuel, (2) mass loss caused by chemical interactions and vaporization, and (3) brittle failure caused by thermal stress. The temperature limitation is probably the preferential loss of constituents from the fuel and the resulting decrease in melting point. The short operating times of nuclear rockets results in compatibility, fission gas release, and changes in chemical and physical properties caused by fission products metals and radiation being of less concern.

Solid solutions of uranium carbide with carbides of the refractory metals zirconium, niobium, hafnium, and tantalum are the more promising high temperature fuels. Table 2 provides the nominal melting points for the important refractory carbides. Small additions of uranium carbide have only a modest effect in depressing the melting point of the refractory carbides. For instance, the uranium carbide and zirconium carbide (ZrC) have a near ideal solid solution. Fig. 3 shows the solid solution for the UC-ZrC pseudobinary solidus-liquid phase diagram. The pseudobinary diagram shown is a projection across the ternary phase diagram from $UC_{1.0}$ to the maximum melting point of ZrC_x, and thereby represents the maximum solidus and liquidus temperatures.

Table 2. Binary carbide melting point maxima.

Compound	Melting Temperature, K
$ZrC_{0.81}$	3693 ± 20
$NbC_{0.85}$	3871 ± 50
$HfC_{0.88}$	4200
$TaC_{0.88}$	4273
$UC_{0.98}$	2798 ± 30
$UC_{1.84}$	2720

Diffusion and vaporization of carbon and uranium from the surface of the fuel changes the composition and hence the melting point. Carbon loss by vaporization from ZrC or NbC is very rapid at high temperatures, but drop quickly with time and approach a relatively constant value once diffusion and vaporization composition rates are matched. For longer lifetimes, because of the loss of carbon, carbide fuel on the carbon-rich side of the solid solution is preferable. The presence of < 20 at. % uranium has only a small effect on the congruently vaporizing composition because metal diffusion rates are slow. With the uranium content of the surface is low, the surface of uncoated fuel will have a melting point that is higher than the bulk material.

Fuel or coating mass loss to the hydrogen gas is a function of temperature (Fig. 4). Below temperatures of ~ 1,500 K, an insignificant reaction is expected. Between ~ 1,500 and ~ 2,800 K, hydrogen corrosion occurs. When hydrocarbons are added to the propellant gas, or are acquired by hydrogen corrosion upstream, the mass loss rate is reduced. In spite of the corrosion reaction, and physical changes in the mid-region, mass-loss and physical changes are expected to be small. Hydrogen reaction peak and become negligible at higher temperatures. Above 2,900 K, in the hot end region, mass-loss caused by simple vaporization of the component elements in the fuel becomes significant. The mass loss depends on the operating time, the maximum surface temperature, and the surface area of the fuel exposed to the high temperature. High fuel density is necessary for high performance nuclear rockets. Porosity enhances vaporization and surface diffusion, which in turn limits lifetime. Fig. 5 demonstrates these effects when comparing a NERVA fuel element with a particle bed element.

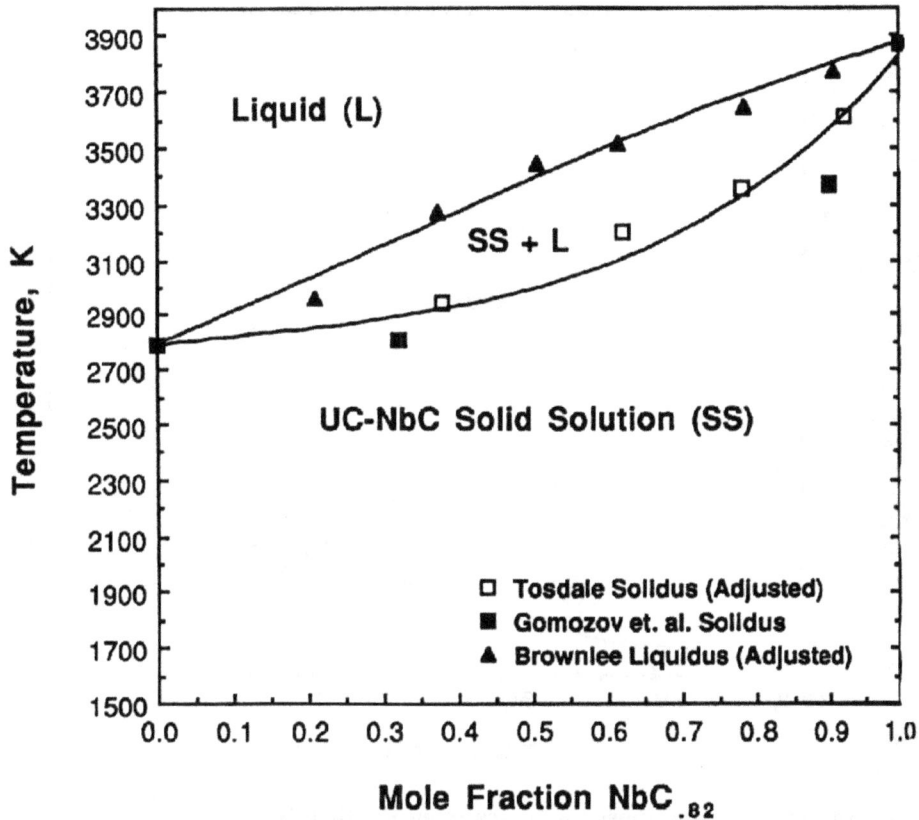

Fig. 3. The UC-ZrC$_{0.31}$ pseudobinary phase diagram showing solidus and liquidus temperature data for the $(U_yZr_{1-y})C_x$ solid solution. *From* **R. Bruce Matthews, et al., 1994.**

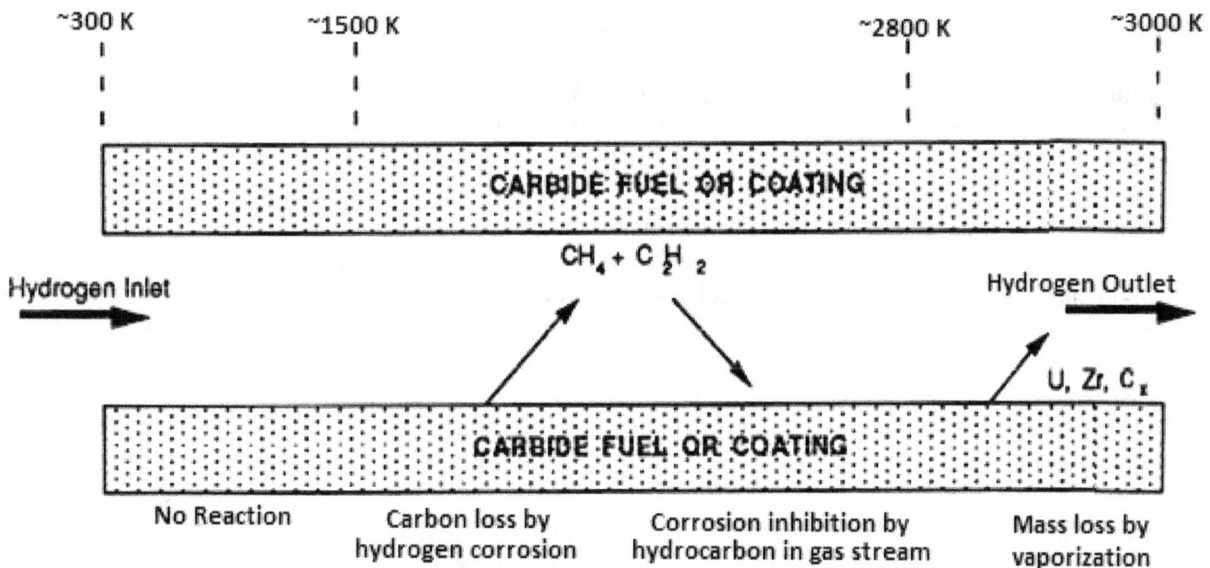

Fig. 4. Schematic representation of the reactions occurring in carbide fueled nuclear rocket core. *From* **R. Bruce Matthews, et al., 1994.**

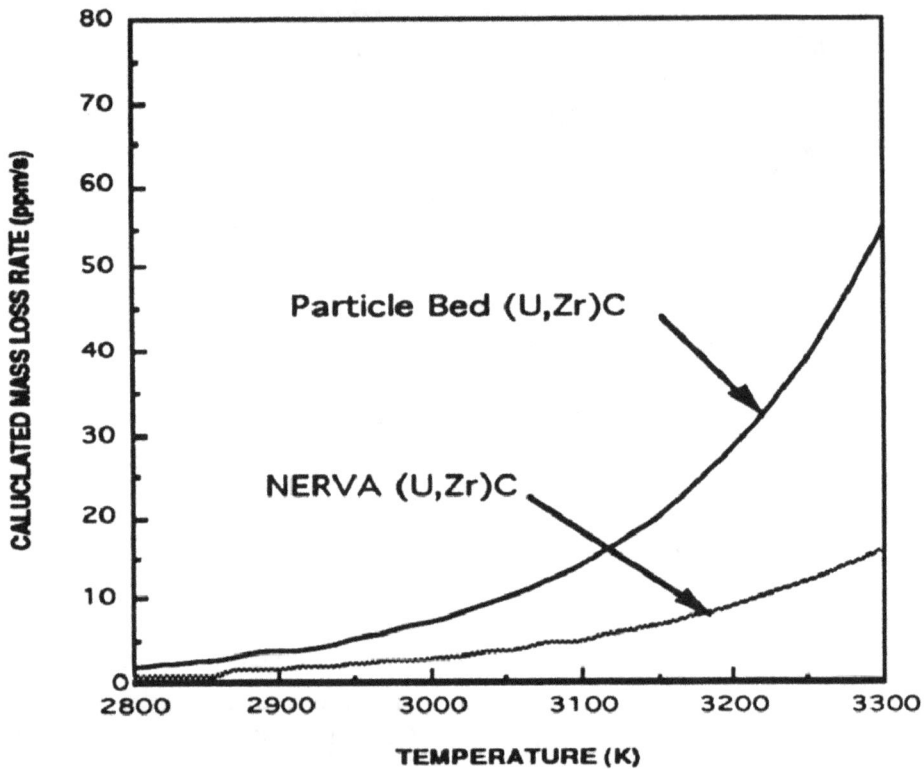

Fig. 5. Comparison of predicted mass loss rates from Particle Bed and Prismatic Nuclear Rocket Core configurations. *From* **R. Bruce Matthews, et al., 1994.**

The practical operating conditions of the fuel is determined by either: (1) the time and temperature that causes enough fuel mass loss or structural deterioration to cause reactor shutdown, or (2) the time and temperature at which the fuel surface composition has changed to the composition of the solidus. Liquid will form on the evaporating surface and destructive processes will be accelerated. Notice that the melting point does not determine the maximum operating time and temperature.

The maximum temperature for both ZrC and NbC based UC fuels mass-loss is predicted to be in the 3,100-3,200 K range for reactors operating < 30 minutes. Considerable additional data will be required to verify carbide fuel performance.

A combination of thermal shock and thermal expansion can lead to mechanical failure of carbide fuels and coatings. Inherent thermal shock resistance of brittle materials is related to the temperature gradient thermal transient and mechanical properties of the fuel. All the refractory carbides have relatively similar thermal conductivities and mechanical properties. Mechanical failure will be significantly affected by fuel element design. The coated Rover fuel element suffered from thermal expansion mismatch between the graphite and ZrC coating, causing cracking of the protective coating. The particle bed design could be affected by thermal expansion causing the bed to expand and either fracture coating materials or distort the containment. Thermal shock and brittle fracture could be the most important limit for the Russian twisted fuel element design.

To recapture the previous fuels work and to develop higher performance fuels, a list of needs and facility requirements are summarized in Table 3.[2] They include fuel fabrication and assembly, non-nuclear materials testing, hydrogen capability testing, fuel capsule testing in a reactor, transient reactor testing, nuclear furnace to duplicate operating conditions of a full scale NTP reactor, and hot cells for evaluating fuel behavior after operating in a radiation environment

Table 3. Fuel development needs and facilities necessary in their development.

FACILITY REQUIREMENTS	MISSION GOALS/OBJECTIVES
Fuel Fabrication and Assembly Category I SNM Facility capable of processing 200 kg U and 1000 fuel elements per year: feedstock preparation; powder preparation; sphere fabrication; sintering, CVD coating; extrusion; hot pressing; graphitizing; brazing; electron beam, laser, and GTA welding; assembly lines; inspections;quality assurance; scrap recovery; and waste treatment.	• Recapture fabrication procedures • Determine phase equilibrium and melting points • Develop new fuels and fuel forms • Develop new fabrication procedures • Fabricate test fuels and fuel elements • Develop fuel element joining techniques • Pilot plant fabrication of test cores • Develop spent fuel recovery procedures • Demonstrate quality-assured procedures
Ex-Pile Testing and Characterization Lab Adjunct to the Fuel Fabrication facility: analytical chemistry, ceramography, NDE, mechanical testing, high temperature testing, H2 testing, compatiblity testing, and kinetic, physical, and thermodynamic properties.	Quantitatively understand: • Thermal transport of material • Thermal stability of fuels and coatings • Chemical stability of fuels and coatings • Thermal stress resistance • Thermal properties for design • Component compatibility • Mass-loss and degradation caused by H2 reactions. • Thermal transient response
Hot Gas Testing Lab Capable of heating NTP fuel elements to 3500 K in flowing hydrogen, with data collection and analysis, post-test characterization, hydrogen and SNM containment	Quantitatively understand: • Corrosion mechanisms • Hydrogen compatibility at high gas flow rates • Coating integrity and stability at high gas flow rates • Fuel and coating mass loss at high gas flow rates
Capsule/Test Reactor Small test reactor with instrumented capsules, fuel temperatures to 3500 K for 10 hours, in hydrogen atmospheres, NDE equipment, data collection and analysis, in-line fission gas analysis	• Screening of solid solution fuel forms Quantitatively understand: • Fission product release • Hydrogen compatibility • Irradiation induced swelling • Compatibility with fission products
Transient Test Reactor Rapid thermal transient testing of fuel elements and assemblies	• Restart and cycling capability • Thermal stress resistance • Off-normal operation • Fission product release
Nuclear Furnace Able to duplicate operating conditions of a full scale NTP reactor with data collection and analysis, fission product containment, and prototype gas flow rate	• Restart and cycling capability • Element/element interactions • Corrosion mechanics • Statistical irradiation data
Hot Cells Burnup analysis, neutron radiography, profilometry, gamma scan, ceramograpy, fission gas analysis, SEM, microprobe, analytical chemistry.	• Postirradiation examination of tests for fission gas release, swelling, mass loss, compatibility, etc.

Engine Cycle Design Considerations

The engine cycle selection impacts the performance, reliability, development risk, and time and cost of engine system development. It directly affects the specific impulse, thrust-to-weight ratio, chamber pressure and pressure levels throughout the engine, restartability, transit operations including startup, shutdown and throttleability, reliability and complexity. Candidate nuclear thermal propulsion cycles are mainly: expander (or topping), gas generator and hot bleed cycles (see Fig. 9).[3] In the expander cycle, the turbine is powered from propellant extracted prior to the reactor core and the turbine exhaust is reinserted above the reactor core. All of the propellant is heated in the core and exhausted through the nozzle at maximum core exit temperatures. Energy to drive the turbine comes from the nozzle, reflector and structures. The gas generator cycle uses a separate gas generator to drive the turbine. This means that an additional propellant, probably liquid oxygen, is added to the system. Gas from the generator is expelled to space through a separate nozzle. The bleed cycle extracts propellant from the nozzle through a port and mixes this with cooler propellant to drive the turbine. The propellant from the turbine is exhausted to space through a separate nozzle.

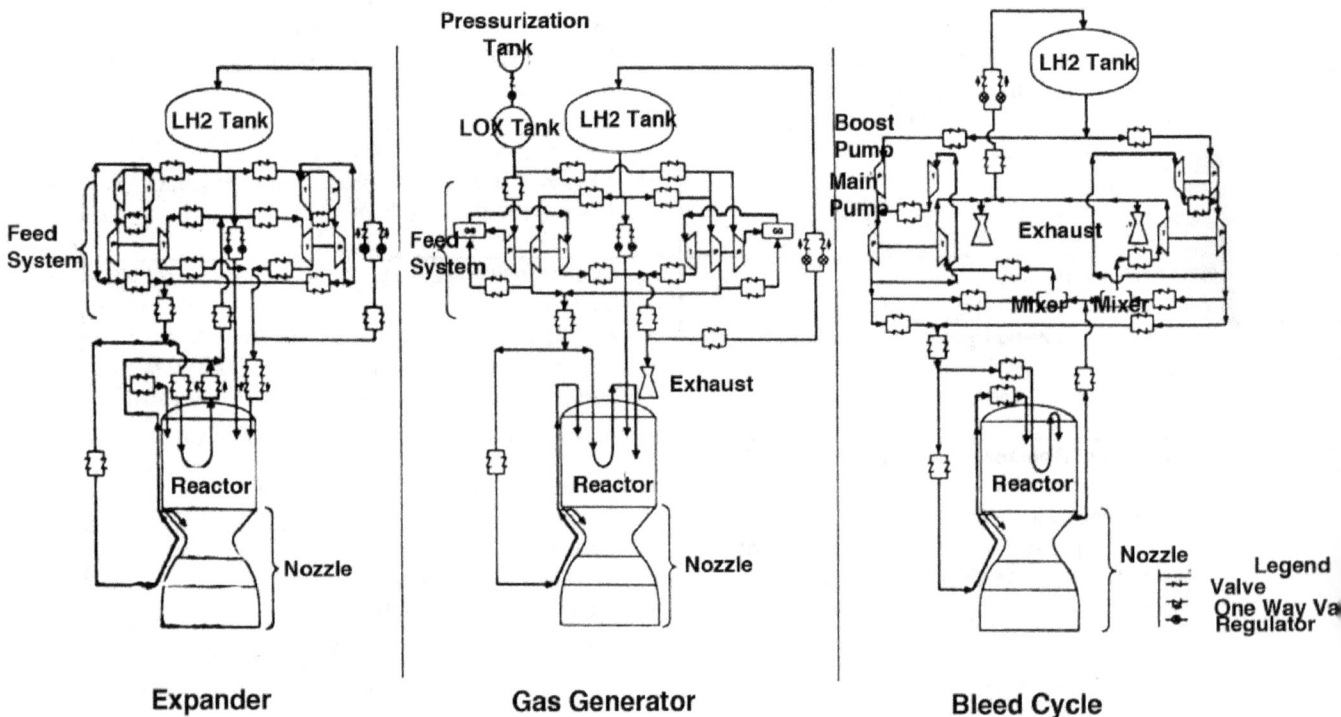

Fig. 9. Representative NTP engine cycles. *From David Buden, Lawrence R. Redd, Timothy S. Olson, Robert Zubrin, 1993.*

The expander cycle has the lowest life cycle cost with its higher performance and lower weight. It operates at low to modest chamber pressures, but is typically a complex, highly integrated cycle design. Because all of the exhaust propellant is hydrogen and exits the engine at the core exit temperature, there is no penalty from lower performance propellant exiting through separate nozzles as with the gas-generator or hot bleed cycle. This cycle was the choice for the NERVA flight engine design.

The gas generator NTP engine cycle options represents a simpler, decoupled cycle design. It has lower performance and higher weight than the other cycle choices. Development risk are expected to be lower than for the expander or hot bleed cycles because it is the cycle used in chemical rockets and can be developed in parallel with other components. However, it introduces the need for another propellant, probably liquid oxygen, and thus adds complexity to the vehicle design. The cycle is easily scalable. With the performance penalties and the need for LOX, this cycle has not to date been selected for NTP engines.

The bleed cycle extracts heated propellant from the nozzle, mixes it with cooler fluid upstream from the reactor core to the desired turbine operating temperature, and exhaust the propellant out of the turbine to space through a separate nozzle. The disadvantage of this cycle is the lower performance compared to the expander cycle as a result of the turbine exhaust being at lower temperatures then the core exit temperature. However, the performance is better than the gas generator cycle. This cycle was used in the XE Experimental Engine and was demonstrated in an engineering prototype of a flight system.

Mission Success And Crew Safety Considerations For Mars Missions[4]

Performing a Mars mission successfully and safely involves getting the crew there and back in good physical and mental condition or, in case of a malfunction, to return the crew safely back to Earth. Nuclear thermal propulsion was selected as the reference propulsion system for manned Mars missions because of its high specific impulse. This results in faster trip times for all opportunities with much reduced propellant requirements, minimizes spacecraft assembly needed in Earth orbit, and increases windows of opportunity for going to and from Mars. The implications are: reduced crew exposure to galactic radiation, reduced psychological and physiological stresses to the crew, less chance of leaving Earth orbit for Mars without a spacecraft thoroughly checked out, and less chance of troubles in assembling the spacecraft prior to leaving Earth orbit. Most of the safety issues have already been addressed regarding mission and crew safety; here we will concentrate on issues associated with establishing the thrust level and the number of rocket engines.

A number of arrangements are considered, including both single and multiple engines, and considered with and without redundant active elements. Fig. 10 shows a one engine configuration from the NERVA development program that included redundant turbopumps and valves. This configuration required 28 valves in the system to isolate a failed turbopump or valve. With a single failure of an active element, the engine would continue to operate. The NERVA engine was being designed for a mission reliability of 0.995. Triple redundancy was analyzed; however, the passive elements were found to dominate the reliability, and the engine reliability was not any higher than that for dual redundancy.

Fig. 11 shows a configuration with three engines, but no active element redundancy within a given engine. There are redundant valves from the tank to avoid loss of propellant from a leaking valve. It is interesting to notice that the three engines in total have 30 valves that include triple redundancy for the cooldown CSCV valves. Thus, a three engine configuration has basically the same number of valves as a single engine with redundant turbopumps and valves. Single engine reliability is 0.95. Fig. 12 shows the effect of engine number on mission success and safe return of the crew. NERVA engine reliability values were used in the analysis. These were then degraded by a factor of ten to see the effect of 0.95 mission reliability for each engine.

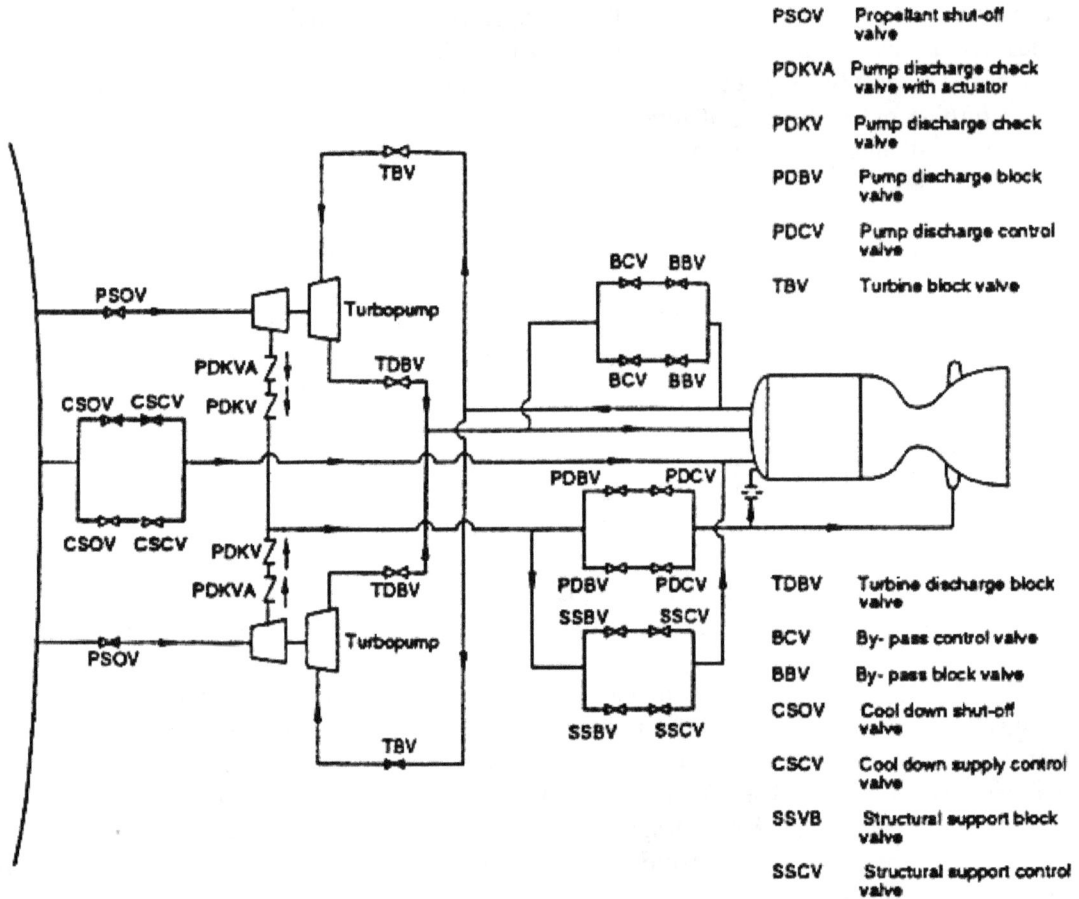

PSOV	Propellant shut-off valve
PDKVA	Pump discharge check valve with actuator
PDKV	Pump discharge check valve
PDBV	Pump discharge block valve
PDCV	Pump discharge control valve
TBV	Turbine block valve
TDBV	Turbine discharge block valve
BCV	By-pass control valve
BBV	By-pass block valve
CSOV	Cool down shut-off valve
CSCV	Cool down supply control valve
SSVB	Structural support block valve
SSCV	Structural support control valve

Fig.10 Single engine reliability drives towards redundant active components (single engine-fail safe). *From David Buden, Lawrence R. Redd, Timothy S. Olson, Robert Zubrin, 1993.*

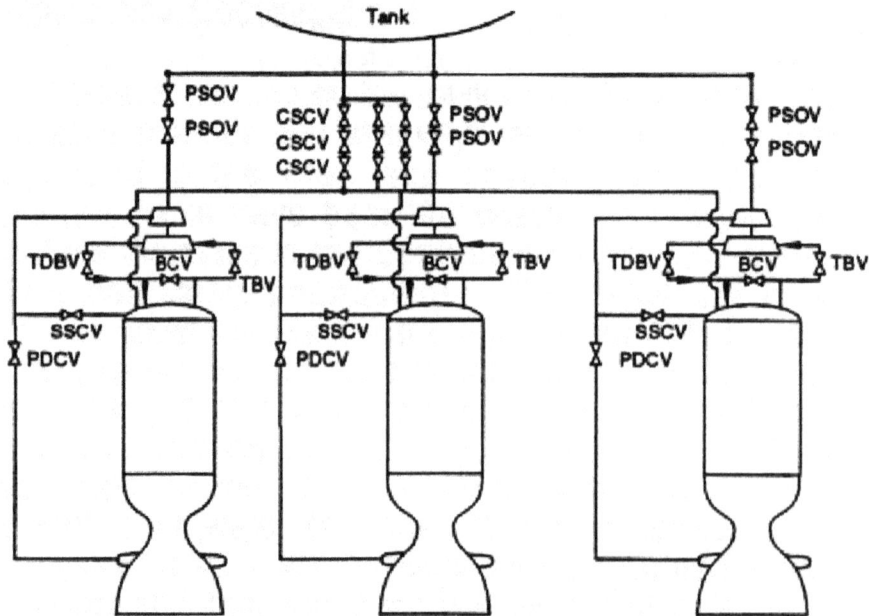

Fig. 11. Clustered engines drives toward simpler engines and high propulsion module reliability (three clustered engines each designed to fail operational/fail safe). *From David Buden, Lawrence R. Redd, Timothy S. Olson, Robert Zubrin, 1993.*

Common cause and non independent failures are also considered. Common cause failures relate to conditional failures given a failure of one component where identical components are present elsewhere in the propulsion module. The commercial nuclear power plant industry has found that other models of that component will fail at an increased rate. Generic values used are to increase the failure rate for the second component by 0.1; and for the third by 0.5. However, these values can be reduced by strict quality control standards, independent inspection of all operations, and detailed understanding of the designs. These standards apply to space systems and certainly must be instituted for the propulsion module. The common cause values can then be reduced to about 0.1 and 0.3 respectively. If strict quality standards are not applied, redundancy buys very little; with strict standards, redundancy significantly increases the propulsion module reliability (based on the same failure rate for each element).

A non-independent failure is a failure that causes one or more additional failures. These occur only in the small percentage of the time that an engine failure might occur. Non-independent failures that are catastrophic to the mission/crew are much less likely for nuclear rockets than occur with chemical rockets. This is a result of having only a monopropellant present and less energetic turbomachinery. The most significant non-independent failure of concern is the destruction of the turbopump, with energy on the order of 10,000 hp at design operations. This energy could be released by unloading the pump, perhaps by excess operation at a critical speed or by a high head, low flow condition resulting in vapor generation along the impeller. The turbopump needs to be designed and oriented to auger away from other engines and propellant tanks. Another potential non-independent failure sometimes considered is a meltdown of one reactor in a redundant propulsion module. Experience in the ROVER program, where a complete loss of flow occurred, was that the reactor containment vessel remained intact. In addition, an emergency cooldown tank, pressurized off of the pump discharge pressure, can be used to handle both emergency and normal cooldown situations.

Notes and assumptions:
- Fail safe condition: crew return with no further backup(s)
- Mission success: 1 failure away from "fail safe"
- Engines are considered credible failures
- No common cause or non-independent failures considered

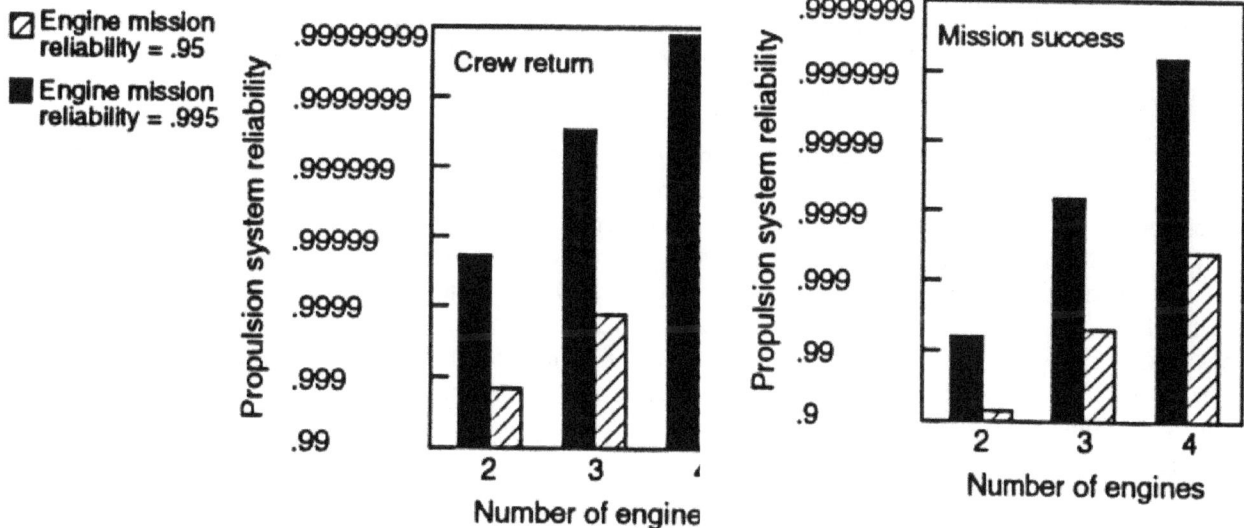

Fig.12. **Safety and mission success favor multiple engines.** *From David Buden, Lawrence R. Redd, Timothy S. Olson, Robert Zubrin, 1993.*

Interactions between clustered engines include neutronic coupling, shutdown core power generation, shutdown reactivity, and reactor kinetics and control. Experiments, called KIWI-PARKA, were run in the 1960's on coupling at Los Alamos National Laboratory and calculation models were used to predict the extent of the reactor coupling (see Fig. 13).[5] The results show that the coupling reactivity between cores is small and is inversely proportional to separation distance and to core size. For example, with a 10 m stage and four engines located at a radius of 4 m, the reactivity interaction is less than $0.1. This is small compared to the several dollars of reactivity in the control drums.

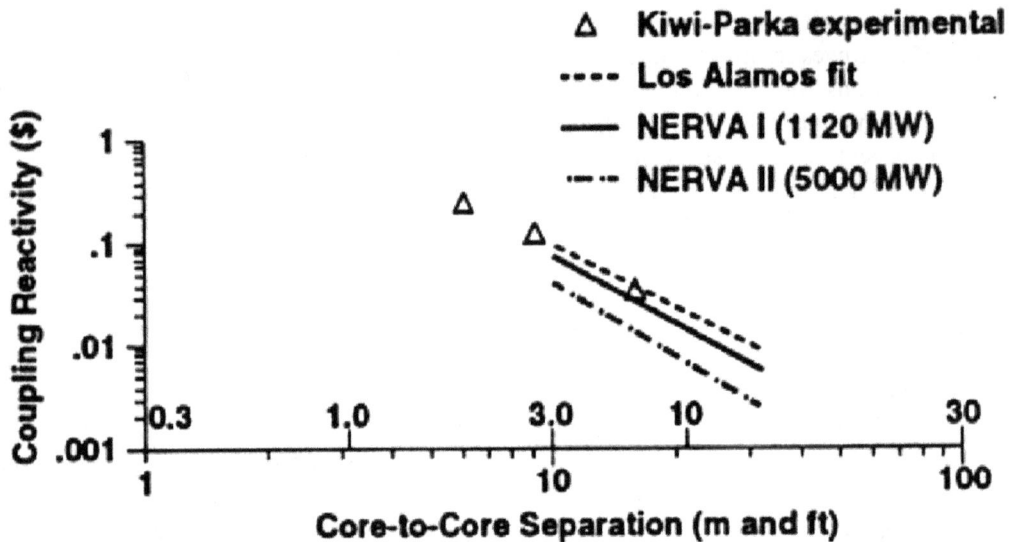

Fig. 13. Experimental and analytical predictions of coupling reactivity as a function of core separation. *From* **John J. Buksa, et al., 1992.**

One unwanted consequence of neutronic coupling is the subcritical fission power generated in an engine out situation. Even though the shutdown reactor is far from critical, the nearby operating engines act as a relatively significant source of neutrons, which can undergo multiplication and thus produce power. The driver core leakage term scales linearly with power level and the geometric term between cores varies little for constant core separation distances. The cluster shutdown engine heat generation rates resulting from decay power and neutronic coupling to adjacent operating engines is such that the pulse cooling decay heat system should be adequate to handle this.[6] The coupling power is less than the decay heat power. Reactor kinetics and control are not major problems. The control systems will probably use temperature control with overrides to protect against excessive fuel temperatures.

Another issue concerns mission operations. Operations must be performed in a safe and reliable manner and strongly determines the safety provided to the crew and environment. Multiple perigee kicks have been advocated as a means to reduce the mass in low Earth orbit for low total thrust. This has safety implications. First of all, the crew makes additional passes through the Van Allen belts. The effects from this were quantified and found to add less than 0.2 rem out of some 40 rem projected from natural radiation in a 150 day trip to Mars. This assumes that the crew is shielded by the equivalent of 6 inches of water (an amount that is conservative compared to what will be needed for a solar flare storm shelter) and that the crew will be in the solar flare storm shelter during the few hours of nuclear propulsion operations. The total time in the Van Allen belts is: 30 minutes for a single burn Trans Mars Injection, 2 hours for a two burn Trans Mars Injection, and 4 hours for a three bum Trans Mars Injection.

Another safety consideration of multiple perigee burns is the time available to assess the performance of the entire spacecraft before the final commitment to leave for the multiyear Mars mission. With a single burn, the time is about one hour, but with three perigee burns there is about 20 hours to assess the functionality of all the onboard systems. This time can be extended by allowing extra orbits between burns for checking equipment.

Thrust misalignment is also an operational concern. However, studies show that the thrust misalignment would need to go undetected for very long periods of time to become a significant hazard to the Earth. For example, a five minute bum at perigee with a 50 degree misalignment of thrust from the propulsion module would result in only a 30 km loss of perigee altitude. Therefore, thrust misalignment is not a safety problem.

Midcourse corrections require only a small amount of thrust. This can be accurately accomplished by operating only one engine. At Mars, capture can be achieved with only two out of four engines operational. This provides great flexibility and reliability in safely completing this part of the mission.

Though not in the current plan, if for some reason the mission must be aborted, a powered flyby of Mars can be performed and the spacecraft can be placed on an Earth return path with less than full thrust, specific impulse or propellant mass.

A nuclear rocket propulsion module that can operate in degraded modes, that is one with less than full thrust, specific impulse, or even with some loss of propellant, significantly adds to the safe return of the crew to Earth. Trans-Earth insertion can be achieved with degraded operations of both specific impulse and thrust. Regarding thrust, only one out of four engines is needed to return home. Degraded performance was evaluated assuming the trip times back to Earth are increased or more propellant is carried. The penalties (see Fig. 14) are quite small, with a ten day trip extension with the loss of three out of four engines. Thus, the probability of successful return even in a degraded condition is very high.

Disposal can be performed following the final midcourse correction several days out from Earth. The vehicle is already out of the plane of the ecliptic, and thus is aligned to miss Earth due to the return from Mars' plane. A mean time between encounter of the spent stage and the Earth has been calculated to be approximately 10 million years with non propulsive disposal.[7] Fission product decay occurs in about 300 years. The conclusion from propulsive disposal of engines was that the penalty of additional propellant mass of at least half the engine mass did not warrant the additional non-encounter extension.

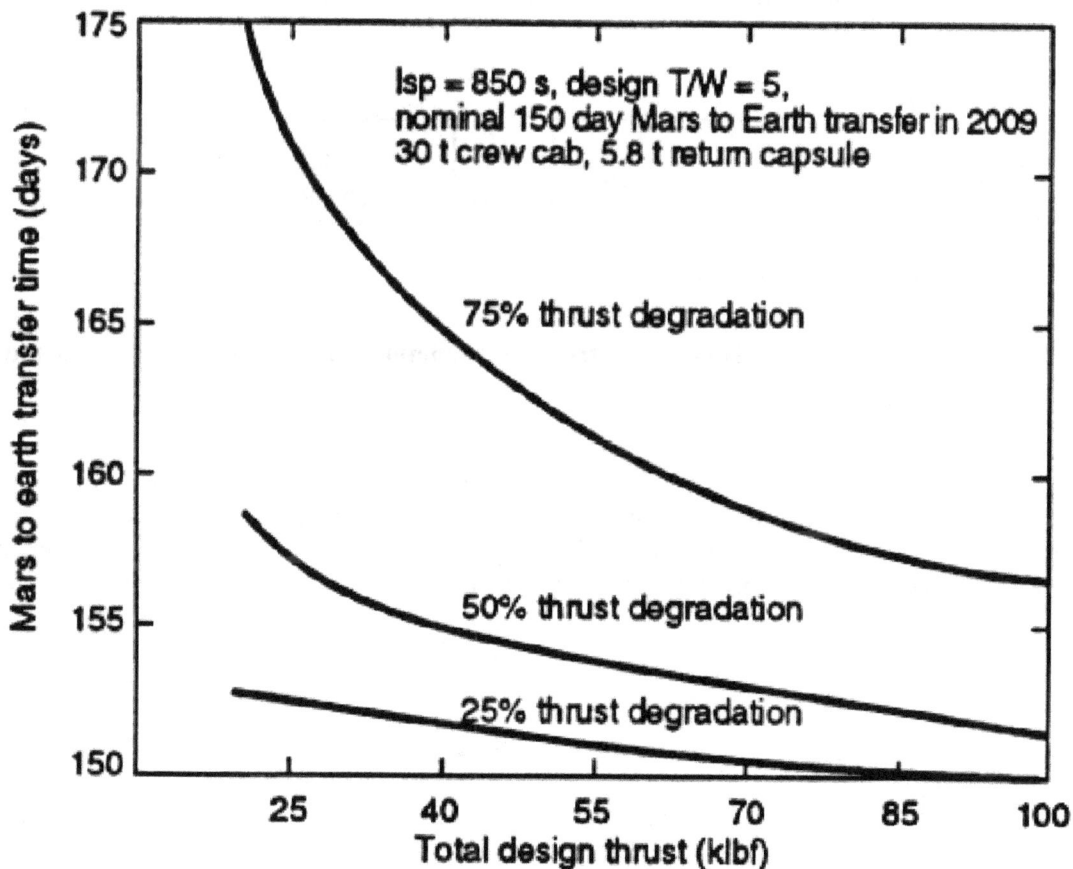

Fig. 14. Increased return transfer time from thrust degradation

Meeting More Restrictive Engine Testing Environmental Standards

Except for the last test series in the Rover/NERVA program (Nuclear Furnace 1), test reactors and rocket engine hydrogen gas was released directly into the atmosphere. This was without removal of the associated fission products and other radioactive debris. With the remoteness of the Nevada Test Site, this was acceptable. However, current rules for nuclear facilities are far more protective of the general environment and potential hazardous quantities of radioactive waste into the atmosphere must be scrupulously avoided.

Several approaches to fission-product retention and waste handling are possible including: (1) closed cycles that involve venting the exhaust to a closed volume or recirculating the hydrogen in a closed loop; or (2) to open cycle where in real time the effluent is processed and vented. As the hydrogen coolant flows through the reactor core, a certain amount of contamination with fission products and/or fuel particulate is possible. The effluent treatment system of this exhaust involves four basic functions (see Fig. 15):[8]

- Initial cooling of the hot reactor exhaust temperatures (range of 2,400 to 3,400 K) to temperatures compatible with normal engineering materials. Also, any debris and large particulate ejected from the core must be retained and maintained in a subcritical configuration.
- Intermediate cooling to temperatures at or below atmospheric. This is considered a desirable part of the design process.
- Fission-product retention to prevent uncontrolled release of contaminants to the environment. This stage retains small particulates, halogens, noble gases, and other volatile species.
- Waste stream processing to properly handle retained fission products, cleaned or process hydrogen effluent, and other potentially contaminated fluids introduced in or generated by the system.

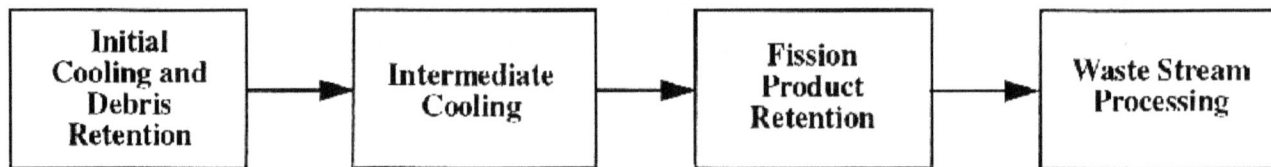

Fig. 15. Reactor and engine testing effluent treatment functions. *From* **Larry R. Shipers and John E. Brockmann, 1993.**

To illustrate how test facilities might be configured for effluent treatment systems, we used 445 kN (100,000 lbs) thrust with a 6.89 MPa (1,000 psia) reactor exit pressure, a test-nozzle area ratio up to 25 to 1, reactor exit temperature of 2,556 K (4,600 R), and nozzle exit stagnation temperature of 2,500 K Fig. 16 is an open cycle with clean hydrogen being flared; Fig. 17 is a closed cycle with the hydrogen being condensed for reuse.[9]

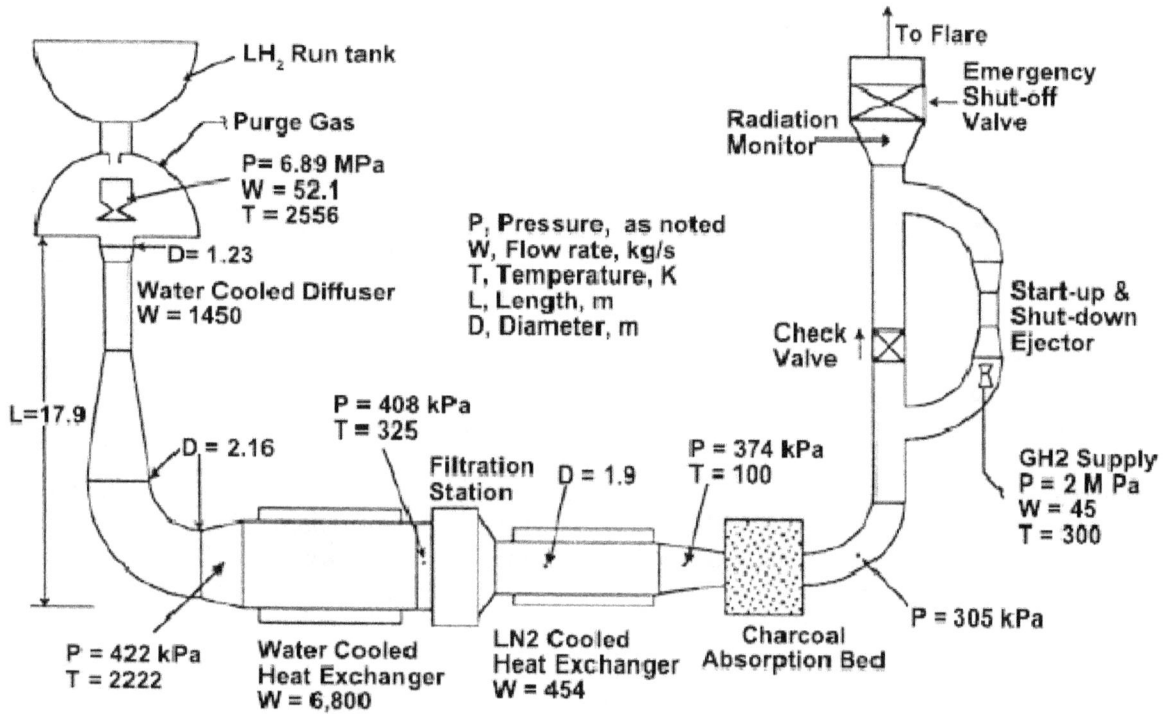

Fig. 16. Nuclear Thermal Rocket Test Facility, open cycle effluent system. *From* **Larry R. Shipers and John E. Brockmann, 1993.**

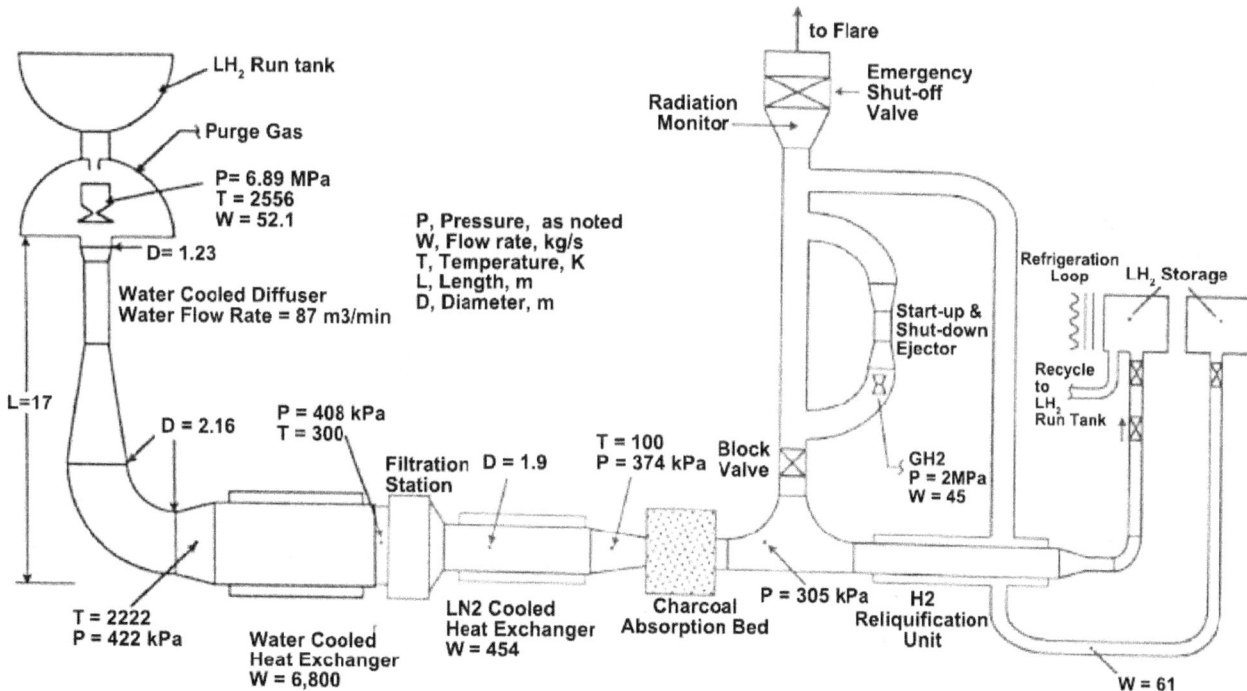

Fig. 17. Nuclear Thermal Rocket Test Facility, closed loop system. *From* **Larry R. Shipers and John E. Brockmann, 1993.**

At an area ratio of 2, the nozzle exit static pressure of about 13 kPa (1.9 psia) is significantly below atmospheric pressure. The rocket exhaust must be pumped to flow against atmospheric back pressure, and the surrounding containment pressure must be kept at or below 13 kPa to permit the nozzle to flow full. The nozzle exhaust flowing into a diffuser provides the required jet pumping, except during startup and shutdown. An auxiliary ejector is required for startup and shutdown. The diffuser from the engine nozzle exit plane is about 14.7 m in length and might be made of 3.2 mm OFHC copper tubing.

The diffuser for a downward-firing engine would transition to the horizontal from the exit of the subsonic diffuser using a long-radius copper tubing elbow. The exit of the elbow transitions to the first-stage, water-cooled, heat exchanger. A debris trap is located in this area, upstream of the heat exchanger, to collect solid particles from the rocket engine, and to protect the downstream heat exchanger tubing and other equipment from physical damage. The heat exchanger would drop the temperature to about 325 K and thus required about 6,800 kg / s water flow rate. The heat exchanger design could be either tubes in a parallel or cross-flow arrangement with an effective length of 21.3 m and 4,655 m^2 transfer area.

The temperature of the hydrogen gas needs to be reduced further to about 100 K to facilitate removal of krypton and xenon by activated charcoal. The heat exchanger could be similar to the water-cooled unit, but use liquid nitrogen as the coolant. The spent nitrogen could be used for purging other system parts prior to venting to the atmosphere. The nitrogen would be monitored for radioactivity in case of leakage before venting.

An activated charcoal bed is used to remove noble gas fission products. The bed operates at 100 K to enhance its effectiveness in removing the xenon and krypton.

Effluent downstream of the charcoal beds is monitored for residual activity prior to release for burning, re-compression and storage, or liquefaction of the hydrogen. If activity levels are above permissible limits, the test would be terminated and remedial action taken.

For closed cycles, one might consider building a huge cavern underground to store the gas and allow it to cool slowly; combustion of the gas with oxygen and condensation of the water for further processing later; or liquefy the hydrogen for further use.

From the experience in Nuclear Furnace I and design studies of effluent treatment system[10, 11, 12] the stricter environmental hazards standards for radioactive waste into the atmosphere can be met.

Overall Relative Maturity of NTP Designs

The Rover/NERVA program successfully demonstrated with real hardware the development and extensive proof testing of nuclear thermal propulsion technology. This enables NASA to plan to use NTP in future missions with a high degree of confidence. Milton Klein provided some key points in summarizing the Rover/NERVA program that are worth emphasizing.[13]

> *"As one would expect, there are some development process truisms that are encountered in an extensive program such as this.*
>
> *Despite knowledge of underlying science and extensive engineering design and analytical efforts, there will be surprises that are only found by full system testing.*
>
> *But full system tests are costly in time and money. So, before a full system test it's worth trying to find any misunderstandings and the unexpected by extensive component and subsystem testing.*

Although understanding the science is essential, we found that sometimes answers are only found through experience-some stuff is "black-magic."

Mission planners must be involved in setting development objectives.

For major development efforts, there is the issue of how far to proceed with major work before a mission is approved in order that mission planning can be done with confidence in the technology to be employed. Although this program was terminated when the classes of missions for which it was intended were put off to the indefinite future, NASA now has a technology that it can count on."

Today, only the Rover/NERVA engine technology can be considered sufficiently technological mature having demonstrated in engineering test to be considered ready for flight development. The Russian NTP propulsion is close, but lacks an engineering prototype test. Other concepts have been proposed with some very attractive features: such as, higher power densities; improved specific impulse; ease of fabrication and testing; designs that reduce or eliminate the need for propellant to remove decay heat; and increased safety in the design, such as fueling the reactor in space. Though these other concepts may incorporate one or more of these features, none of these other concepts has reached a sufficient level of experimental verification and engineering development to be considered ready for flight development.

Some Day -- Fusion

To obtain a quantum jump in propulsive capabilities in the next 30 to 50 years, designs should be based on technologies that are well understood today. Fusion, as a source of propulsion, is limited by the fact that we are still far from demonstrating the technology for commercial power, let alone space applications. Many conceptual studies exist on fusion systems that indicate order of magnitude gains possible. These include both inertial and magnetic fusion concepts. These devices will continue to evolve, but are at least 50 years into the future.

Anti-matter is another potential form of order-of-magnitude gains in propulsive technology. However, it is severely hampered by extremely low propulsion efficiencies and the high costs of current production methods.

Currently, the only feasible concept using fusion energy is one that derives thrust from plasma waves generated from a series of small, supercritical fission/fusion pulses behind an object in space. For spacecraft applications, a momentum transfer mechanism translates the intense plasma wave energy into a vehicle acceleration that is tolerable to the rest of the spacecraft and its crew. A program to do this (Project ORION) existed between 1958 and 1965.[14] The ORION design, shown in Fig. 18, had a diameter of 10 to 30 meters, with high thrusts of 1 to 10 g accelerations and high specific impulse of approximately 10,000 s. When the Partial Test Ban Treaty in 1965 outlawed the use of nuclear explosives in space, the project was terminated.

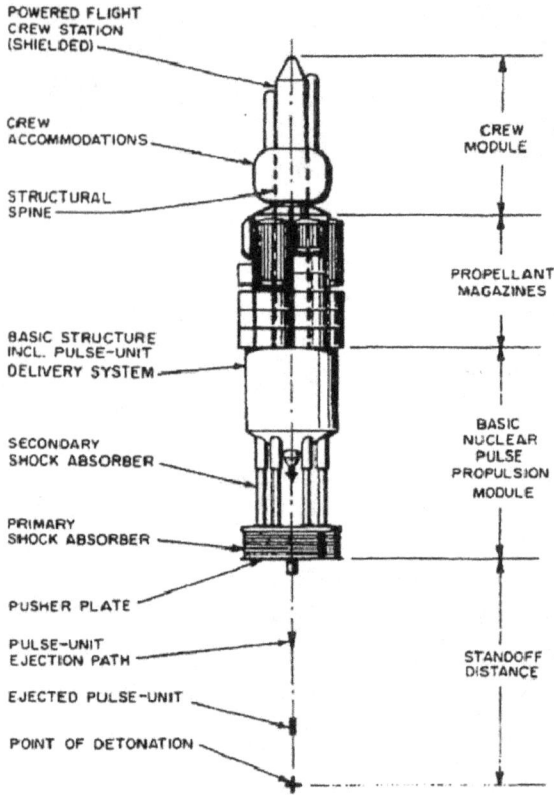

Fig.18 . 1960 Project ORION spacecraft concept. *From J. A. Bonometti, P. J. Morton, and G. R. Schmidt, 2000.*

NOTES

OVERVIEW

[1] Dennis Miotla, "Assessment of Plutonium-238 Production Alternatives," Briefing for Nuclear Energy Advisory Committee, April 21, 2008.

[2] H, M. Dieckamp, *Nuclear Space Power Systems,* Atomics International, Canoga Park, CA, September 1967

[3] Atomic Industrial Forum, *Guidebook for the Application of Space Nuclear Power Systems,* New York, January 1969.

[4] D. Buden et al., "Selection of Power Plant Elements for Future Reactor Space Electric Power Systems," LA-7858, Los Alamos National Laboratory, NM, September 1979.

[5] Ibid, D. Buden, et al.

[6] U.S. Department of Energy, "Environmental Development Plan (EDP)--Space Applications," DOE/EDP-0026, April 1978.

[7] "Nuclear Energy in Space," DOE Office of Nuclear Energy, Science & Technology, DOE/NE0071.

[8] Bennett, Gary, et al, "Mission of Daring: The General-Purpose Heat Source Radioisotope Thermoelectric Generator," 4th International Energy Conversion Engineering Conference, AIAA 2006-4096, June 2006.

[9] Jet Propulsion Laboratory Internet Site, www.jpl.nasa.gov.

[10] U. of Wisconsin lectures, fti.neep.wisc.edu/neep602/SPRING00/lecture34.pdf

[11] U.S. Department of Energy, ebid.

CHAPTER ONE

[1] J. M. Taub, "A Review of Fuel Element Development for Nuclear Rocket Engines," *Los Alamos National Laboratory LA-5931*, 1974.

[2] Joseph A. Angelo, Jr. and David Buden, *Space Nuclear Power,* Orbit Book Company, 1985.

[3] John R. Lamarsh, *Introduction to Nuclear Engineering* (2nd Edition), Addison-Wesley Publishing Co., Reading, Mass., 1983.

[4] S. Glasstone and M. C. Edlund, *The Elements of Nuclear Reactor Theory*, VanNostrand Co. Inc., Princeton, NJ, 1952.

[5] John R. Lamarsh, *Introduction to Nuclear Reactor Theory*, Addison-Wesley Publishing Co., Reading, Mass., 1972.

[6] D. Buden et al., "Selection of Power Plant Elements for Future Reactor Space Electric Power Supplies," LA-7858, Los Alamos Scientific Laboratory, September 1979.

[7] Atomic Industrial Forum, "Guidebook for the Application of Space Nuclear Power Systems," January 1969.

[8] H. M. Dieckamp, "Nuclear Space Power Systems," Atomics International, September 1967.

[9] J. H. Van Osdol et al., "Design and Integration of Zirconium Hydride Reactor Power Systems," AI-AEC-13072, Atomics International, June 1973.

[10] P. N. Stevens and H. C. Claiborne, *Weapons Radiation Shielding Handbook*, Chapter 2, DASA-1892-5, Defense Atomic Support Agency, June 1970.

[11] A. M. Weinberg and E. P. Wigner, *The Physical Theory of Nuclear Chain Reactors*, University of Chicago Press, 1958.

[12] Argonne National Laboratory, "Reactor Physics Constants", ANL- 5800 (2nd edition), July 1963.

[13] Joseph A. Angelo, Jr. and David Buden, *Space Nuclear Power,* Orbit Book Company, Malabar, FL, 1985.

CHAPTER TWO

[1] *America At The Threshold, America Space Exploration Initiative*, The Synthesis Group Report, 1991.

[2] David Buden, "A Development Approach For Nuclear Thermal Propulsion," *NTSE-92 Nuclear Technologies For Space Exploration, An American Nuclear Society Meeting*, Jackson, Wyoming, Aug. 1992, pp. 350-360.

[3] Larry Redd, Tim Olson, Warren Madsen, "Design-of-Parameters Approach to Nuclear Thermal Propulsion Systems Analysis," *NTSE-92 Nuclear Technologies for Space Exploration, An American Nuclear Society Meeting*, Jackson, Wyoming, 1992, pp. 860-869.

[4] Stanley K. Borowski, "The Rational/Benefits of Nuclear Thermal Rocket Propulsion for NASA's Lunar Space Transportation System," Paper AIAA 91-2052, 1991.

[5] Stanley K. Borowski, et al., "Nuclear Thermal Rocket/Stage Technology Options for NASA's Future Human Exploration Missions to the Moon and Mars," *CONF 940101, 1994 American Institute of Physics, AIP Conference Proceedings 301*, Eleventh Symposium Space Nuclear Power and Propulsion, Albuquerque, NM, 1994, pp. 745-758.

[6] Frank E. Rom, "Review of Nuclear Rocket Research At NASA's Lewis Research Center From 1953 Thru 1973," *Paper AIAA 91-3500*, 1991.

[7] Synthesis Group Report, America At The Threshold, America's Space Exploration Initiative, available from the Superintendent of Documents, U.S. Government Printing Office, Washington, D.C., 20402, June 1991.

[8] Buden, David, "Safety Questions Relevant to Nuclear Thermal Propulsion," 9th symposium On Space Nuclear Power Systems, Albuquerque, New Mexico, January 1992.

[9] Larry R. Shipers and John E. Brockmann, "Effluent Treatment Options For Nuclear Thermal Propulsion System Ground Tests," *CONF 930103, 1993 American Institute of Physics*, pp. 1005-1016.

[10] Gary L. Bennett, et al, "Prelude to the Future: A Brief History of Nuclear Thermal Propulsion in the United States," in *A Critical Review of Space Nuclear Power and Propulsion 1984 - 1993*, Editor Mohamed S. El-Genk, 1994 American Institute of Physics, L.C. Catalog Card No. 94-70780, New York, pp221-267.

[11] "Project Pyro, Liquid Propellant Explosive Hazards", Final Report, AFPRO-TR-68-92, December 1968.

[12] Bennett, G. L., "Safety Status of Space Radioisotope and Reactor Power Sources," 25th Intersociety Energy Conversion Engineering Conference, Reno, Nevada, August 1990.

[13] Bennett, G. L., "Flight Safety Review Process For Space Nuclear Power Sources," 22nd Intersociety Energy Conversion Engineering Conference, Philadelphia, Pennsylvania, August 1987.

[14] Lee, J., et al, "Technology Requirements for the Disposal of Space Nuclear Power Sources and Implications for Space Debris Management Strategies," AIAA/NASA/DOD Orbital Debris Conference: Technical Issued and Future Directions, *Paper No. AIAA 90-1368*, Baltimore, Maryland, April 16-19, 1990.

[15] Angelo, J., and D. Buden, "Post Operational Disposal of Space Nuclear Reactors," 21st Intersociety Energy Conversion Engineering Conference, San Diego, California, August 25-29, 1986.

[16] Milton Klein, "Nuclear Thermal Rocket -- An Established Space Propulsion Technology," *CP699, Space Technology and Applications International Forum--STAIF 2994*, edited by M.S. El-Genk, 2004 American Institute of Physics 0-7354-0171-3/04, pp. 413-419.

[17] Buden, D., L. R. Redd, T. S. Olson, R. Zubrin, "NTP Design Specifications for a Broad Range of Applications," Paper No., AIAA 93-1947, AIAA/SAE/ASME/ASEE 29th Joint Propulsion Conference, June 28-30, 1993, Monterey, CA.

CHAPTER THREE

[1] Richard J. Bohl, et al., "History of some Direct Nuclear Propulsion Developments Since 1946," *Space Nuclear Power Systems 1987*, Orbit Book Co., Malabar, FL., 1988, pp. 467-473.

[2] Milton Klein, "Nuclear Thermal Rocket -- An Established Space Propulsion Technology," *CP699, Space Technology and Applications International Forum--STAIF 2994*, edited by M.S. El-Genk, 2004 American Institute of Physics 0-7354-0171-3/04, pp. 413-419.

[3] *G. H. Farbman and R. E. Thompson, "Applications of Nuclear Rocket Technology to Light Weight Nuclear Propulsion and Commercial Nuclear Process Heat Systems," AIAA Paper No. 75-1261.*

[4] *David S. Gabriel, "Nuclear Propulsion in the United States," 1972.*
[5] *Westinghouse Astronuclear Laboratory, "NRX-A6 Test Predictions, "WANL 1613, November 1967.*
[6] Daniel R. Koenig, "Experience Gained from the Space Nuclear Rocket Program (Rover)," Los Alamos National Laboratory *Report LA-10062-H, UC-33,* May 1986.
[7] *Luther L. Lyon, "Performance of (U, Zr)C-Graphite (Composite) and (f (U, Zr)C(Carbide) Fuel Elements in the Nuclear Furnace 1 Test Reactor," Los Alamos National Laboratory report LA-5398-MS, September 1973.*
[8] *David S. Gabriel's Statement to Committee on Aeronautics and Space Services, U. S. Senate, 1973.*
[9] *F. D. Durham, LA-5044-MS, Los Alamos National Laboratory.*
[10] W. H. Robbins and H. B. Finger, "An Historical Perspective of the NERVA Rocket Engine Technology Program," *AIAA Paper 91-3451.*
[11] David Buden, "Operational Characteristics of Nuclear Rockets," *AIAA 5th Propulsion Joint Specialist Conference, U. S. Air Force Academy,* Colorado, AIAA Paper No. 69-515, 1969.
[12] Buden, D., "Nuclear Rocket Safety," 38th Congress of The International Astronautical Federation, Paper No. IAF-87-297, Brighton, United Kingdom, October 1987.
[13] AGC, "NERVA Safety Plan," Document No. S-019-22-090205-F1, Aerojet Nuclear Systems Co., September 1970.
[14] AGC, "Flight Safety Contingency Analysis Report", Document No S-103-CP090290-F1, Aerojet Nuclear Systems Co., September 1970
[15] AGC, September 1970.
[16] *L. A. Booth and J. H. Altseimer, "Summary of Nuclear Engine Dual-Mode Electrical Power System Preliminary Study," Los Alamos National Laboratory Report LA-DC-72-1111.*
[17] Stanley K. Borowski, "The Rational/Benefits Of Nuclear Thermal Rocket Propulsion For NASA's Lunar Space Transportation System," *AIAA Paper 91-2052,* 1991.
[18] James T. Walton, "An Overview of Tested and Analyzed NTP Concepts," *AIAA Paper No. 91-3503.*
[19] Lyman J. Petrosky, "ENABLER-II: A High Performance Prismatic Fuel Nuclear Rocket Engine," *CPMF 920104,* 1992 American Institute of Physics, pp. 728-737.
[20] ibid Lyman J. Petrosky

CHAPTER FOUR

[1] Joseph R. Wetch, et al., Development of Nuclear Rocket Engines in the USSR," Presented at *AIAA/NASA/OAI Conference on Advance SEI Technologies,* Sept. 4-6, 1991, AIAA Paper 91-3648.
[2] Robert R. Corban, "NTP System Definition and Comparison Process for SEI," *CONF 930103, 1993 American Institute of Physics, AIP Conference Proceedings 271,* Tenth Symposium Space Nuclear Power and Propulsion, Albuquerque, NM, 1993, pp. 1713-1721.
[3] Vadim Zakirov and Vladimir Pavshook, "Russian Nuclear Rocket Engine Design for Mars Exploration," *Tsinghua Science and Technology, ISSN 1007-0214 04/08, Vol. 12, Number 3,* June 2007.
[4] Ibid Corban
[5] ibid Zakirov and Pavshook.
[6] David L. Black and Stanley V. Gunn, "Space Nuclear Propulsion," *Encyclopedia of Physical Science and Technology, Third Edition, Volume 15, 2002 Academic Press,* pp. 555 - 575.
[7] A. Borisov, " Russian Nuclear Rocket Engines," Historical records, Cosmonautics News, January 2001 (in Russian).
[8] ibid Wetch, et al.
[9] N. N. Ponomarev-Stepnoy, et al, "Space Nuclear Power and Propulsion Plants Based on Solid Core Nuclear Reactor with External Power Conversion Unit," International Conference on Nuclear Energy in Space--2005, Russia, March 1-3, 2005 (in Russian).
[10] R. Bruce Matthews, et al., "Fuels For Space Nuclear Power And Propulsion: 1983-1993," *A Critical Review of Space Nuclear Power And Propulsion 1984-1993,* Editor Mohamed S. El-Genk, L. C. Catalog Card No. 94-70780, 1994 American Institute of Physics, pp.179-210.

CHAPTER FIVE

[1] F. L. Horn, J. R. Powell, and O. W. Lazareth, "Particle Bed Reactor Propulsion Vehicle Performance and Characteristics as an Orbital Transfer Rocket," *Space Nuclear Power Systems 1986*, Edited by M. S. El-Genk and M. D. Hoover, Orbit Book Co., Malabar, FL., 1987, pp. 375-381.
[2] Paul S. Ma, Lewis A. Walton, and Mathew W. Ales, "Particle Contact Forces In Particle Bed Reactor Fuel Elements," *CONF 940101, 1994 American Institute of Physics*, pp. 213-223.
[3] J. R. Powell, et al., "Nuclear Propulsion Systems for Orbit Transfer based on the Particle Bed Reactor," *Space Nuclear Power Systems 1988, Edited by M. S. El-Genk and M. D. Hoover*, Orbit Book Co., Malabar, FL, 1989, pp. 185-198.
[4] "Timberwind," *FAS Nuclear Resources*.
[5] Russell R. Jensen, et al., "Evolution of Particle Bed Reactor Fuel," *CONF 940101, 1994 American Institute of Physics*, pp. 195-204.
[6] ibid Russell R. Jensen,
[7] Jonathan K. Witter, David D. Lanning, and John E. Meyer, "Flow Stability Analysis Of A Particle Bed Reactor Fuel Element," *CONF 930103, 1993 American Institute of Physics*, pp. 1541-1546.
[8] DeWayne L. Husser, et al., "Fuel Design For Particle-Bed Reactors For Thermal Propulsion Applications," *CONF 940101, 1994 American Institute of Physics*, pp. 205-212.
[9] ibid. Jonathan K. Witter

CHAPTER SIX

[1] Mohamed S. El-Genk and Nicholas J. Morley, "Pellet Bed Reactor For Nuclear Thermal Propelled Vehicles," *CONF-910116, 1991 American Institute of Physics*, pp. 607-617.
[2] Nicholas J. Morley and Mohamed S. El-Genk, "Thermal-Hydraulics Sensitivity Analysis Of The Pellet Bed Reactor For Nuclear Thermal Propulsion," *CONF 940101, 1994 American Institute of Physics*, pp. 773-781.
[3] Richard B. Harty and Robert G. Brengle, "Wire Core Reactor For Nuclear Thermal Propulsion," *CONF 930103, 1993 American Institute of Physics*, pp. 571-577.
[4] Lyman J. Petrosky, et al. "IMPULSE - An Advanced, High Performance Nuclear Thermal Propulsion System," *CONF 930103, 1993 American Institute of Physics*, pp. 565-569.
[5] Rolv Hundal, et al., "An Evaluation Of The IMPULSE NTP System Core Configuration," *CONF 940101, 1994 American Institute of Physics*, pp. 750-764.
[6] Carl F. Leyse, et al., "Pressure Fed Nuclear Thermal Rockets For Space Missions," *Seventh Symposium on Space Nuclear Power systems, CONF-900109*, Albuquerque, NM, January 1990, pp. 53-60.
[7] Carl F. Leyse, et al., "Space Nuclear Propulsion - The Low Pressure Nuclear Thermal Rocket." AIAA 90-1952, July 1990.
[8] John S. Clark, "A Comparison Of Nuclear Thermal Propulsion Concepts: Results Of A Workshop," *CONF-910116, 1991 American Institute of Physics*, pp. 740-747.
[9] J. Wesley Davis, et al. "Advanced NTR Options," CONF-910116, *1991 American Institute of Physics*, pp. 748-853
[10] Frank E. Rom, "Review of Nuclear Research At NASA's Lewis Research Center From 1953 Thru 1973," AIAA 91-3500.
[11] George Maise, et al., "The Liquid Annular Reactor System (LARS) Propulsion," *CONF-910116, 1991 American Institute of Physics*, pp. 618-624.
[12] Ibid. Frank E. Rom, 1991.
[13] Ibid. Frank E. Rom, 1991.
[14] K. Thom, "High Grade Power from Fissioning Gases," NASA Report, 1976.
[15] K. Thorn and F. C. Schwenk, "Gaseous Fuel Reactor Systems for Aerospace Applications," *AIAA Conference on the Future of Aerospace Power Systems, AIAA Paper No. 77-513*, March 1977.
[16] Ibid, K. Thom and F. C. Schwenk.

[17] Thomas S. Latham and Claude Russel Joyner II, "Summary of Nuclear Light Bulb Development Status," *AIAA Paper 91-3512*.

[18] Stanley Chow, "Mini-Cavity Plasma Core Reactors for Dual Mode Space Power Propulsion Systems," *The Princeton University Conference: Partially Ionized and Uranium Plasmas*, September 1976, pp. 217-223.

[19] H. Weinstein, "Review of Coaxial Flow Gas Core Nuclear Rocket Fluid Mechanics," *The Princeton University Conference: Partially Ionized and Uranium Plasmas*, September 1976, pp. 123-129.

[20] L. L. Lowry, "Gas Core Reactor Power Plants Designed for Low Proliferation Potential," *Los Alamos National Laboratory Report No. LA-6900-MS*, September 1977.

[21] D. M. Barton et a!., "Characteristic of Vortex Contained UF_6 Gas in a Beryllium Reflected Reactor," *1980 Annual American Nuclear Society Meeting*, Las Vegas, Nev., 8-13 June 1980.

[22] Jay F. Kunze and Albert G. Gu, "Open Cycle Gas Core Startup and Operational Stability," *NTSE-92 Nuclear Technologies For Space Exploration, American Nuclear Society Meeting*, Jackson, Wyoming, 1992, pp. 372-379.

[23] Nils J. Diaz, et al. "Gas Core Reactor Concepts And Technology Issues And Baseline Strategy," *AIAA Paper 91-3582*.

[24] Edward T. Dugan, Yoichi Watanabe, Stephen A. Kuras, Isaac Maya, Nils J. Diaz, "Nuclear Design of a Vapor Core Reactor for Space Nuclear Propulsion," CONF 930103 @ 1993 American Institute of Physics, pp.655- 662.

CHAPTER SEVEN

[1] R. Bruce Matthews, et al., "Fuels For Space Nuclear Power And Propulsion: 1983-1993," *A Critical Review of Space Nuclear Power And Propulsion 1984-1993*, Editor Mohamed S. El-Genk, 1994 American Institute of Physics, L C. Catalog Card No. 94-70780, pp. 179-220.

[2] S. K. Bhattacharyya, et al., "Space Exploration Initiative Fuels, Materials and Related Nuclear Propulsion Technologies Panel," *NASA Technical Memorandum 105706*, September 1993.

[3] Dennis G. Pelaccio, Christine M. Scheil and John T. Collins, "Engine Cycle Design Considerations For Nuclear Thermal Propulsion Systems," *CONF 930103, 1993 American Institute of Physics*, pp. 1729-1736.

[4] David Buden, Lawrence R. Redd, Timothy S. Olson, Robert Zubrin, "NTP Design Specifications For A Broad Range Of Applications," *AIAA Paper 93-1947, AIAA/SAE/ASME/ASEE 29th Joint Propulsion Conference*, Monterey, CA, June 1993.

[5] John J. Buksa, et al., "Nuclear Thermal Rocket Clustering 1: A Summary of Previous Work and Relevant Issues," *CONF 920104, 1992 American Institute of Physics*, pp. 1089-1102.

[6] Ibid. John J. Buksa, etal.

[7] M. L. Stancati and A. L. Friedlander, "Disposal Modes for Mars Transfer Nuclear Propulsion," *AIAA Paper 01-3410, AIAA/NASA/OAI Conference on Advanced SEI Technologies*, Cleveland, OH, September 1991.

[8] Larry R. Shipers and John E. Brockmann, "Effluent Treatment Options For Nuclear Thermal Propulsion System Ground Tests," *CONF 930103, 1993 American Institute of Physics*, pp. 1005-1016.

[9] Herbert R. Zweig, et al., "Exhaust Gas Treatment In Testing Nuclear Rocket Engines," *CONF 930103, 1993 American Institute of Physics*, pp. 999-1003.

[10] Ibid. Herbert R. Zweg, et al.

[11] ibid Larry R. Shipers and John E. Brockmann

[12] Thomas J. Hill, et al., "Space Nuclear Thermal Propulsion Test Facilities Accommodation At INEL," *CONF 930103, 1993 American Institute of Physics*, pp. 1017-1021.

[13] Milton Klein, "Nuclear Thermal Rocket -- An Established Space Propulsion Technology," *CP699, space Technology and Applications International Forum--STAIF 2004*, edited by M. S. El-Genk, 2004 American Institute of Physics 0-7354-0171-3/04, pp. 413-419.

[14] J. A. Bonometti, P. J. Morton, and G. R. Schmidt, "External Pulsed Plasma Propulsion And Its Potential For The Near Future," *CP504, Space Technology and Applications International Forum--2000*, edited by M. S. El-Genk, 2000 American Institute of Physics 1-56396-919-X, pp. 1236-1241.

www.ingramcontent.com/pod-product-compliance
Lightning Source LLC
Chambersburg PA
CBHW081542220326
41598CB00036B/6532